Coastal Research Library

Volume 11

Series Editor

Charles W. Finkl
Department of Geosciences
Florida Atlantic University
Boca Raton, FL 33431
USA

The aim of this book series is to disseminate information to the coastal research community. The Series covers all aspects of coastal research including but not limited to relevant aspects of geological sciences, biology (incl. ecology and coastal marine ecosystems), geomorphol-ogy (physical geography), climate, littoral oceanography, coastal hydraulics, environmental (resource) management, engineering, and remote sensing. Policy, coastal law, and relevant issues such as conflict resolution and risk management would also be covered by the Series. The scope of the Series is broad and with a unique crossdisciplinary nature. The Series would tend to focus on topics that are of current interest and which carry some import as opposed to traditional titles that are esoteric and non-controversial. Monographs as well as contributed volumes are welcomed.

More information about this series at http://www.springer.com/series/8795

Gautam Kumar Das

Estuarine Morphodynamics of the Sunderbans

Springer

Gautam Kumar Das
Department of Chemical Engineering
Jadavpur University
Kolkata, India

ISSN 2211-0577 ISSN 2211-0585 (electronic)
ISBN 978-3-319-36343-1 ISBN 978-3-319-11343-2 (eBook)
DOI 10.1007/978-3-319-11343-2
Springer Cham Heidelberg New York Dordrecht London

Softcover reprint of the hardcover 1st edition 2015

Printed on acid-free paper

Springer is part of Springer Science+Business Media (www.springer.com)

. . . All things rush on, they stop not, they look not behind, no power can hold them back, they rush on.

Rabindranath Tagore (1912)
Gitanjali (Hymn No. LXX)

Dedicated to the People of Sunderbans

Foreword

The Sunderbans, a UNESCO World Heritage Site covering parts of Bangladesh and the southern tip of Indian state of West Bengal, is a part of the world's largest deltaic plain of fluvio-marine deposit formed by the Ganges and Brahmaputra at the confluence of the Bay of Bengal. It is the largest single block of tidal halophytic mangrove forest in the world, conspicuous for its great size and biodiversity. With an anastomising network of channels and creeks, and tidal inundation twice daily, Sunderbans mangroves wetland is a dynamic and complex ecosystem, which undergoes continuous processes of erosion and accretion. Natural processes like changes in local hydrology, sediment motion under wind, wave and tidal action, beach dynamics, regional and global processes like sea level rise as well as the impact of human interference in the form of reclamation of forest land, changes in land use patterns, coastal urbanizations etc. are the lead factors for the changes in the environmental scenario of Sunderbans. Only a century ago, Indian Sunderbans with a total area of 9,630 km^2 was covered with lush green mangrove forests, which at present is left with only 4,266.6 km^2 area with mangroves and the rest has been converted to various land use patterns.

Global warming has left its imprint in the form of sea level rise in the Sunderbans. Sagar Island, the biggest deltaic island of the Sunderbans, has recorded 3.1 mm/year of sea level rise (IPCC 2007). As a result of this there has been flooding of low-lying deltas, change of shoreline, and salinisations have been decreasing gradually due to mixing of seawater with melt water of ice. Again, the incursion of flooded coastal saline water into the freshwater region increases the salinity of the latter. Silt particles get mixed with water and the resultant heat of mixing as a result of the kinetic energy from those vortexes may cause the water temperature to be warmer than that of the ambient temperature during winter.

Evidences of erosion followed by differential subsidence and sediment filling in several parts of the Sunderbans can be sited from the terraced estuarine bank pattern, undercutting and collapsing of river banks etc. The estimated rate of erosion, as evidenced from the relics of the sea wall exposed in the intertidal beach zone at Bakkhali and Fraserganj in coastal Sunderbans, is 9.8 m/year during the period

1930–1970 and 8.6 m/year during 1971–1995 which is very high. The northeastern, southeastern and southwestern sides of Sagar Island are facing vigorous erosion due to the concerted acts of various natural processes and anthropogenic activities. The erosion rate from 1996 to 1999 was calculated to be 5.47 m/year. The part of the destabilized coastal dunes have been advancing inland at the rate of 17 m/year. Beach erosion has also been computed from the presence of gastropod species *Amalda ampla* which is considered as a biological indicator.

The highly specialized mangrove ecosystem of Indian Sunderbans basically can be treated as an estuarine ecosystem supporting a large number of aquatic organisms living either entire life cycles within the mangroves or visiting the mangrove swamps and waters for food or to breed. The occurrence of both the flora and fauna in this fragile ecosystem has become vulnerable as a result of changing morphodynamics of this estuarine environment. Expansion of mangrove swamp to the salt marsh is the clear indicator of changing trends of the mangrove habitat. Mangrove extension into the marshy areas and the gradual disappearance of islands emerged in the coastal areas because of the increased tidal amplitudes for climatic changes and sea level rise are highly alarming for the Sunderbans.

A large number of endangered species of fin fishes and shell fishes are suffering a severe depletion in population because of random exploitation in addition to the damage caused for the capture of prawn seeds. Continuous prawn seed collection can decrease the density of pneumatophores and the biomass of epiphytic algae as trampling is a perennial problem which results the habitat structure of mangrove plants as well as aquatic organisms. This, in turn, adds to a tremendous threat to the balance of the coastal ecosystem.

Estuarine morphodynamics are further influenced by the changes in river course, impeded fresh water flow, climate variability and sea level changes. All these deleterious operations are currently continuing in most areas of Sunderbans, which, in turn, during the recent decades, have been bringing about a serious change in the geomorphology and hydrodynamics of the water courses together with an obvious change in the nature of nutrient recycling. As a result of these morphodynamic processes and anthropogenic interferences, exploitations of mangrove vegetations have led to the large scale degradation of mangrove areas. Further, mangroves have also been lost through natural causes including erosion and severe siltation as a result of sedimentation. At present, tiger straying is a common phenomenon. The present study encompasses the areas of the trend of salinity decline, increase of temperature of river waters, non-uniformity of bottom topography, erosion, change of the tidal courses and shapes of islands, increasing rate of sedimentation in the river bed and mid channel bar formations; collapsing of bank materials; and random collection of prawn seeds that results in the ecological imbalance in this unique ecosystem. It is expected that the findings of this study would be a potential contribution to enlist the morphodynamic factors responsible for the changing environmental scenario and to formulate a comprehensive and sustainable management plan for the long-term conservation and protection of the mangrove ecosystem in the Indian Sunderbans.

Kolkata, India Gautam Kumar Das

Acknowledgement

I would like to extend my sincere thanks to Sri S. Rakshit, Former Director, Geological Survey of India; Prof. Asokkumar Bhattacharya, Former Head, Department of Marine Science, University of Calcutta; Prof. A. Mazumder, Director and Head, Department of Water Resource Engineering, Jadavpur University; Prof. S. R. Basu, Former Head, Department of Geography, Calcutta University; Prof. S. Dutta, Pro-VC, Jadavpur University; and Prof. H. R. Betal, Former Head, Department of Geography, Calcutta University, for their constant inspiration, guidance and advice.

The author gratefully acknowledges various helps including laboratory facilities provided by the Director, Marine Wing, Geological Survey of India and Head, Department of Chemical Engineering, Jadavpur University.

I also gratefully acknowledge the helps rendered by the Directors of Zoological Survey of India, Sunderbans Biosphere Reserves and Indian Meteorological Department. Most cordially I acknowledge the co-operation and assistance extended by the various departments of the Government of India and West Bengal, Forest Directorate, Sunderbans Tiger Reserves and Sunderbans Development Board, Government of West Bengal.

I gratefully remember the generous helps and hospitality rendered to me during field work and on-board field surveys in the forest of Sunderbans by many forest rangers, guards, fishermen, honey collectors, tour operators and many islanders, without which it would have been impossible to conduct this work. Rokeya and Titas helped in drawings and corrected the proofs.

Contents

Abbreviations and Units

BOD	Biochemical oxygen demand
COD	Chemical oxygen demand
DO	Dissolved oxygen
ECw	Electrical conductivity of water
ECs	Electrical conductivity of soil
IFAD	International Fund for Agricultural Development
IMD	Indian Meteorological Department
IUCN	International Union for the Conservation of Nature
LOI	Loss on ignition
MAB	Man and biosphere
ROOM	Readily oxidisable organic matter
SBR	Sunderbans Biosphere Reserve
STR	Sunderbans Tiger Reserve
TDS	Total dissolved solids
TSS	Total suspended solids
UNDP	United Nations Development Programme
UNEP	United Nations Environmental Programme
ppt	Parts per thousand (‰)
ppm	Parts per million
dS/m	desi Siemens per meter
1 ppt	2161 dS/m
1 dS/m	1000 EC = 1000 ppm
100 hectares	1 km^2

List of Figures

List of Tables

Chapter 1
Sunderbans: Physical Aspects and Configurations

Abstract Sunderbans forms a part of the world's largest fluvio-marine Ganges – Brahmaputra deltaic plain at the confluence of the Bay of Bengal with the largest block of halophytic mangrove forest. It has passed through a dynamic and complex physiographical and geomorphological history. Prograding in phases towards Bay of Bengal during Plio-Pleistocene inter-glacial period, it has left behind numerous distinctive landforms and islands of various shapes and sizes. The intricate network of rivers, tidal creeks and inlets with dynamic flow pattern accelerates the process of erosion and accretion and make the geomorphic set-up of the area a complex one. The important morphotypes of Sunderbans are beaches and inlets, creeks and estuaries, mudflats, coastal dunes, sand flats and mangrove swamps.

Keywords Sunderbans • Mangroves • Bay of Bengal • Ganga – Brahmaputra delta

The Sunderbans is one of the most attractive and alluring wonderland on this earth which seems to inspire human being for much creativity. For myriad reasons the Sunderbans has a unique place in the history, geography, literature and socio-cultural aspects of eastern India and Bangladesh. Lying at the southern tips of West Bengal where the land meets the sea forming a delta, the Sunderbans is the largest single block of tidal halophytic mangrove forest in the world which is conspicuous for its great size and biodiversity. Forming a part of world's largest fluvio-marine Ganga-Brahmaputra deltaic plain at the confluence of the Bay of Bengal, Sunderbans with its endless diversity of vibrating nature is a suitable habitat of world's famous luxuriant mangrove forest. Though apparently serene, this is a place where natural beauty and terror coexists. It is awe-inspiring abode of estuarine crocodiles, king cobra and other varieties of serpents and the famous Royal Bengal Tiger which turns the nature's beauty harsh and vengeful, making the struggle of men for existence in this unique place as an intense task.

Nature has bestowed Sunderbans with eternal beauty, but according to climatologists this spectacular region is the most vulnerable place for destruction directly caused by climatic changes. As per meteorological records, Sunderbans faces many storms of different intensities during the last couple of centuries, but one of the most destructive one it faced in 2009, consequence of being in the eye

G.K. Das, *Estuarine Morphodynamics of the Sunderbans*,
Coastal Research Library 11, DOI 10.1007/978-3-319-11343-2_1

of the cyclone namely Aila. IPCC in 2007 reported that the rising sea level due to global warming could contribute to destruction of over 75 % of the Sunderbans mangroves by the end of twenty first century. If the prediction comes true, it is frightening to think about the ramifications of the consequences there of. In addition of being a beautiful ecosystem, one of the most crucial functions that Sunderbans play is that of a protective natural buffer against natural calamities for about million population of greater Kolkata. Had there not been any barrier of this kind, cyclones and the resulting floods would have devil's day and would be absolutely devastating for the Kolkata metropolis. The morpho-dynamics of the estuarine environment of the Sunderbans, India is extremely complex with the primary processes like productivity, food chain and food web, reproduction, nutrient recycling, mineralization and others related independent processes where the substrate soils, estuarine brackish waters and the mangrove forest are the major components and man plays the roles as the observer, inhabitant and exploiter of this mangrove ecosystem (Vannucci 1989).

1.1 Physical Aspects

The area of Indian Sunderbans is 9,630 km^2 (Table 1.1) out of total area of 25,500 km^2 and the remaining part covers the area of Bangladesh Sunderbans. Sunderbans is famous for its mangrove ecosystem and about 7 % of the world's mangrove falls under India and the area are estimated to be 6,740 km^2. Out of this Sunderbans of West Bengal has the largest area, which is nearly about 4,200 km^2 (Tables 1.1, 1.2, and 1.3). The entire Sunderbans region is distributed between Sunderbans Tiger Reserve (STR) and 24 Parganas South Forest Division which are geographically separated by the Matla River (Fig. 1.1). The Ministry of

Table 1.1 Indian Sunderbans – a statistical account

1.	Total areas:	9,630 km^2
	(a) Areas of mangrove forests:	4,266 km^2
	(b) Project Tiger:	2,585 km^2
	(c) Area of Sunderbans National Park:	1,330 km^2
2.	Total number of islands:	102
	(a) Reclaimed islands:	54
	(b) Forestry islands:	48
	(c) Total areas of agricultural lands:	5,430 km^2
	(d) Development blocks:	19
	(e) Total population:	4.2 million
3.	Total no. of river:	31
4.	Total length of embankment:	4,250 km
5.	Area of Sunderban Biosphere Reserve:	9,630 km^2
6.	Area of world heritage site:	9,630 km^2

Table 1.2 Zonation of mangrove forests

	Total mangrove forest area:	4,266 km^2
A.	Area of Sunderbans Tiger Reserve:	2,585 km^2
	(i) Sajnekhali Wildlife Sanctuary:	362 km^2
	(ii) Buffer zone:	893 km^2
	(iii) National Park (core area):	1,330 km^2
B.	Area for South 24 Parganas Forest:	1,678 km^2
	(i) Holiday Island Wildlife Sanctuary:	6 km^2
	(ii) Lothian Island Wildlife Sanctuary:	3,859 km^2
	(iii) Other forest area:	1,634 km^2

Table 1.3 Areawise zonation of Indian Sunderbans

Zone	Area lying between	Forestry (km^2)	Agriculture (km^2)	Restoration zone (km^2)	Total area (km^2)
I	Sagar-Mohisani Ghoramara-Sand Heads etc.	–	75	15	90
II	Mohisani Island and Thakuran River	–	700	200	900
III	Thakuran and Matla River	1,370	200	30	1,600
IV	Core area of biosphere resource				1,692
	North of buffer area	1,692			
V	Buffer zone of Tiger Reserve	893	–	–	893
VI	Settlement area north of forest area of Tiger Reserve	–	4,455	–	4,455
	Total area in km^2	3,955	5,430	245	**9,630**

Environment and Forests, Government of India declared the entire 9,630 km^2 of Sunderbans as the Sunderbans Biosphere Reserve (SBR) in 1989 for conservation, management, research, training and maintaining harmony between man and environment of mangrove forest lands, Sunderbans (Anonymous 2003).

Geologically the area is the result of extensive fluvio-marine deposits of the river Ganges and Bay of Bengal. Several tectonic lineaments across the Bengal coastal sediments are identified from aerial photographs and satellite imageries which can be correlated with the tectonic history of lower Bengal Basin. Lithological logs of the boreholes at Sunderbans include the thick and effective clay blanket (15–75 m) at the top and is underlain by the presence of very coarse sediment containing medium to small (3 m diameter) gravels of rock fragments (fine grained sand stone and silt stone) and quartz, which is generally attributable to Quaternary age (Allison 1998). The character of the sediments of the Bay facing deltaic islands is sandy, silty clay and clay from the south to the north. Constantly built-up Sunderbans with vast new alluvial plain intersected by a large number of tidal rivers and creeks, surrounded by the saline and brackish water has soil, water, air, flora and fauna as its components of vibrating ecosystem. This entire ecosystem may lose its dynamic nature with the change of these components. Having a very high pressure of about 4.2 million population in the 54 reclaimed human habitable

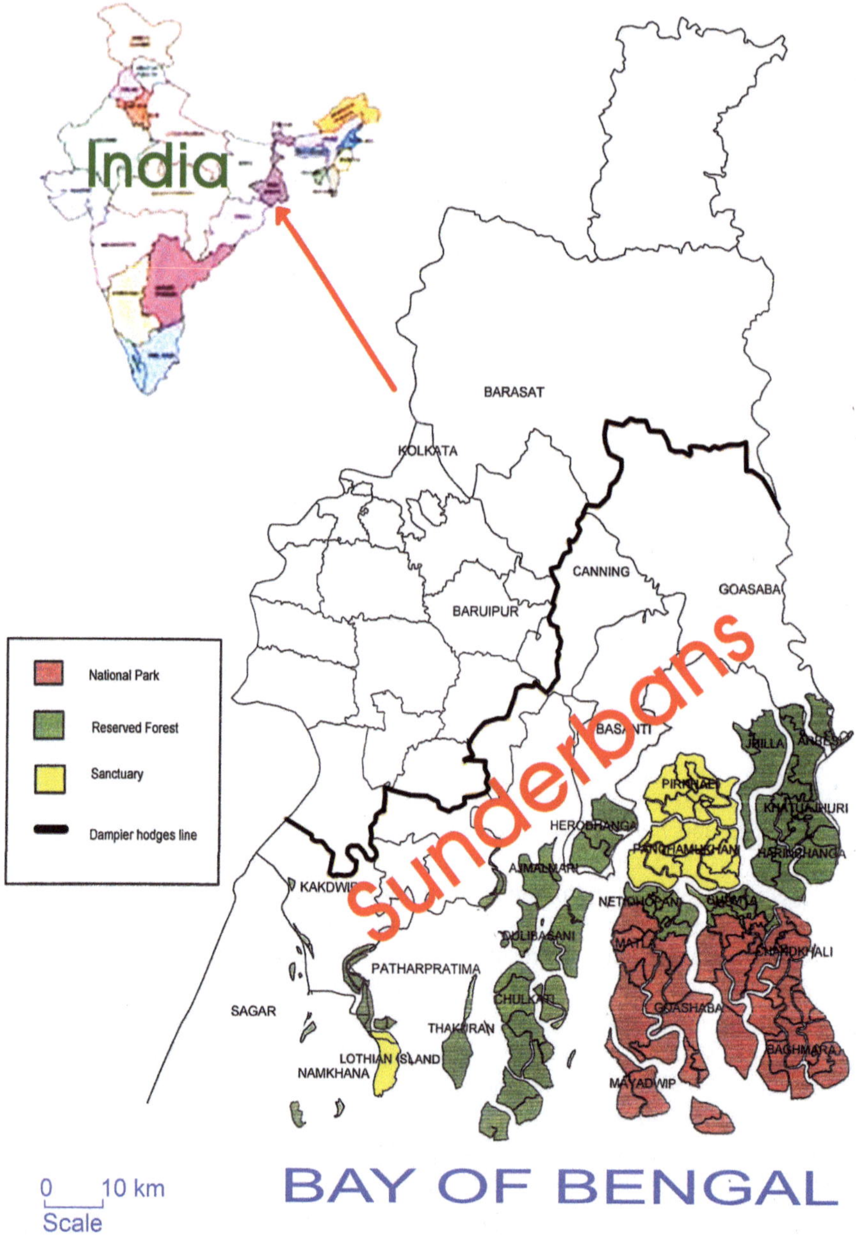

Fig. 1.1 Map of Sunderbans Biosphere Reserves demarcating the forest and reclaimed area bordering with the Dampier Hodges line

islands of its surroundings, Sunderbans always faces threats of destruction. This anthropogenic threats are due to construction of freshwater reservoir for irrigation in the agricultural land or consequent geotectonic movements and tilting of Bengal delta. Total mangrove area of Sunderbans is reduced of late for these reasons. Lack of freshwater discharge from the river upstream causes the loss of mangroves at Sunderbans, as the heads of rivers are no more connected with the river Ganges.

1.2 Location and General Boundaries

Indian Sunderbans forms the largest block of mangroves of the world taken together with Bangladesh (Sharma 2010). It is a part of the world's largest deltaic plain of fluvio-marine deposit formed by the Ganges and the Brahmaputra at the confluence of the Bay of Bengal. The geographical limit of Indian Sunderbans can be demarcated by the Dampier and Hodges line (an imaginary line drawn in 1831) in the north west, the river Hooghly in the west, the rivers Ichamati – Kalindi – Haribhanga in the east which is incidentally the international boundary between India and Bangladesh, and the Bay of Bengal in the south. This stretch is situated within the Latitude 21°31′N and 22°30′N and Longitude between 88°10′E and 89°51′E, encompassing parts of the districts of South 24 Parganas and southern parts of the North 24 Parganas. The total area of Indian Sunderbans is 9,630 km^2 which is nearly one-third of the total 25,500 km^2 areas. Indian Sunderbans has a shore length of 130 km out of the total 180 km coastal length of West Bengal. On the other side, the luxuriant dense mangrove forest at the Bangladesh part of Sunderbans covers about 230 km of shore length of total coastal length of 710 km.

1.3 Reclamation of Sunderbans

Since the eighteenth century, the intertidal dense mangrove forest areas of Sunderbans were reclaimed step by step and converted into agricultural field and brackish water fisheries. More than 55 % areas of the Indian Sunderbans have been cleared or reclaimed till date. The area consists of around 102 islands of which about 54 islands are reclaimed in the north western part for giving way to human habitation (Table 1.1).

1.4 Zonation of Sunderbans

Sunderbans Biosphere Reserve has been classified geographically, into six zones covering an area of 9,630 km^2 (Tables 1.1 and 1.3). There are another three specific zones classified on the basis of conservation of nature namely (i) Core zone (ii) Buffer zone and (iii) Transitional zone (Table 1.2). The transitional zone is comprised of both forestry and agricultural lands. The core zone of the biosphere reserve will be kept absolutely undisturbed. The entire forest area of the buffer zone under the Sunderbans Tiger Reserve serves as the manipulation zone where tourist may visit, fishermen, wood cutters and honey collectors may harvest the forest products through the proper permission of the forest authority. Core area is actually declared as the Sunderbans national Park (Table 1.2). Core areas are no way different from the other part of the Sunderbans other than the zones of no exploitation and activities.

1.5 Climate

The climate of Sunderbans is tropical oceanic. Three seasons viz. winter (November–February), summer (March–June) and monsoon (July–October) are easily recognizable. Winter temperature ranges from 10 to 25 °C and summer temperature from 28 to 36 °C. The annual rainfall ranges between 1,900 and 2,100 mm, Salinity of coastal water ranges from 19 to 31 ‰. pH of coastal water ranges from 7.5 to 8.5. The generally mesotidal coast (tidal amplitude = 2–4 m) is macrotidal (tidal amplitude >4 m) at the funnel mouths of estuaries and big rivers. The tides are semi-diurnal with slight diurnal inequality. The heavy rainfall during the monsoon includes the tidal interactions in almost all the rivers of Sunderbans and flood and ebb tidal currents fluctuate with seasons. The maximum wind velocity is 16.7–20 km/h (April–June) and minimum wind velocity is 10.7–11.8 km/h (December–February). West Bengal is cyclone prone. Number of severe cyclones may even be 3–4 in a year. The wind velocity during cyclones often ranges from 80 to 140 km/h. Cyclones also initiate large-scale littoral drift and lead to devastating coastal modifications. Wave heights range from 0 to 0.6 m with a wave period of 5–7 s during the calm winter season, whereas, these become 1.8–2.4, and 12–14 s respectively during the rough summer seasons. Wave height goes much above 2.5 m during cyclonic storms. The mean wind velocities during the three principal seasons i.e., pre-monsoon, monsoon and post monsoon are 11.5 km h^{-1}, 11.1 km h^{-1} and 6.65 km h^{-1} respectively. The south-southwest to southwest wind direction of pre-monsoon and monsoon changes to north-northeast to northeast during the post – monsoon times. Number of rainy days sharply rises to 65.84 % during monsoon from the pre-monsoon 20.05 % and again declines to 14.11 % during the post – monsoon times.

The mean water temperature decreases with the channel length from the source point towards the river mouth. As the amount of suspended solids decreases and depth of the channel increases, water is observed to retain lesser amount of heat. Water temperature varying between 18.1 and 21.5 °C is well within the range. Mangroves cannot tolerate cold temperature less than 20 °C for a continuous period. For these reasons, mangroves are found only in the tropical and sub-tropical region including Sunderbans. The shallow depth estuarine brackish water temperature is always a few degrees higher than the adjacent coastal waters.

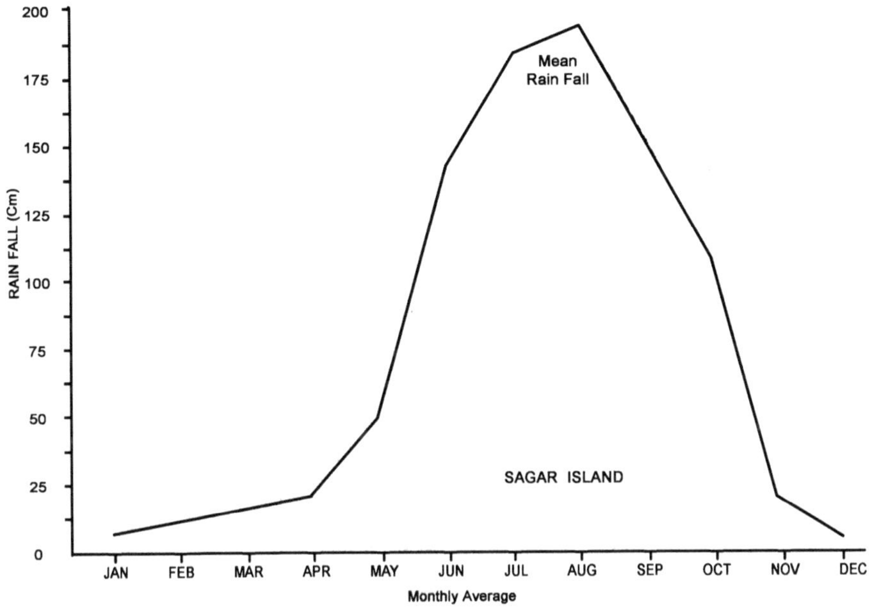

Fig. 1.2 Monthly variation of rainfall (Average of 100 years) at Sagar Island (Data collected from IMD)

Temperature and rainfall data for over a period of 100 years are available from the Indian Meteorological Department (IMD). Rainfall data shows annual mean value of 1,908.4 mm (Fig. 1.2). This becomes slightly higher to the tune of 1949.4 mm during 1991–1993. Average of 3 years (1991–1993) of evaporation data is recorded to be 977 mm and this implies an excess of annual precipitation in the area over evaporation.

1.5.1 Cyclones

Tropical cyclones with variable wind speed (63–87 km h^{-1}) are regular phenomena in the Sunderbans area. Tropical cyclonic depressions with wind speed <63 km h^{-1} frequently pass over the area during July to September. "Cyclonic Storms" with wind speed ranging from 63 to 87 km h^{-1} strike the area during June to September. Most of destructive "severe cyclones" with wind speed >87 km h^{-1} affects the area particularly during May and September. Thus the average recurrence intervals of cyclonic depressions, cyclonic storms and severe cyclonic storms over the study area are twice a year, once every 3 years and again once every 3 years respectively. The 1976 cyclone, that had crossed Sagar Island, on 11th September, is the latest major event till now. Its maximum surface wind speed reached 130–148 km $h^{-}1$ but one of the most destructive one it faced very recent in 2009, consequence of being in the eye of the cyclone namely Aila (Das 2009). This storm caused a wide spread damage in the coastal area of West Bengal, adjoining state Orissa and the neighbouring country Bangladesh.

1.5.2 Tides

The tide in the Sunderbans is semi-diurnal with little diurnal inequality and approximately synchronous along the river courses of the Sunderbans. Tidal currents are the strongest hydrodynamic influence on the sub-aerial delta front (De Boer et al. 1989) and sub-aqueous part of the deltaic Sunderbans. Deformation of tide front entering the islands and channels of the Hugli-Matla estuaries produces tidal amplitudes exceeding 3.4 m and tidal currents up to 290 cm/s in the river mouths in this estuarine delta. The mean flood velocity is 52 cm/s, whereas, the mean ebb velocity is 56 cm/s. Differences in the channel aspect ratio (water depth: width) and tidal asymmetry in the river mouth estuary leads to flood dominance in the eastern tidal channels and ebb dominance in the western. Tidal length in different rivers and inlets of Sunderbans ranges from 60 to 80 km, although in Hugli River it is 290 km and tides are measurable up to that limit. The inland limit of saline influence follows an irregular line west of the Hugli-Matla estuary seaward up to the 3 m elevation contour and depends on the size and separation of the distributary channels that dissect the Ganga- Brahmaputra delta front. The entire coastline of West Bengal with its highly indented nature is the outcome of differential erosion and accretion caused by fluvio-tidal and tidal processes in an estuarine delta framework. The shore perpendicular elongated shape of most of coastal islands reflects the impacts of flood and ebb tidal currents in a macrotidal (Davies 1964) coastal environment. The intertidal landforms also result from high tidal fluctuations with unequal inundations and exposures.

At the sea face the mean tide is 5.5 m whereas, in the middle stretch it declines to 3.5 m. At river mouths, the mean maximum spring tidal range varies between 4.5 and 5.5 m and the mean minimum neap between 2.0 and 2.5 m (Das 2009). The highest yearly tides are experienced during August to September while the lowest during February to March. The mean ebb velocity is 58 cm/s, whereas, the mean flood velocity is 48 cm/s. Time velocity asymmetry controls both bedforms configurations and their orientations on the flood and ebb channels.

1.6 Waves

The northern Bay of Bengal adjacent to the Sunderbans has a moderate wave climate with average wave heights of less than 0.5 m and 3–4 s wave periods. During the monsoon season wave heights average 0.5–1.0 m, with occasional waves up to 2 m with corresponding periods of 6 s. Larger waves up to 5 m have been observed only during cyclones. Tropical cyclones can affect coastal Sunderbans in both the pre monsoon (March–June) and post-monsoon (September–December) period with devastating waves of larger heights. Owing to the low elevation of the coastal plain of deltaic Sunderbans, the resulting storm surges can penetrate as far as 100 km inland and have resulted in the death of millions of persons and loss of household wealth. Wave parameters, micro and macrotidal cycle, long shore currents are recorded in the Sunderbans in the different seasons. Coastal processes are very dynamic and it is induced by the tropical cyclones of the area. The coastline is changing in the reclaimed areas due to the abnormal cliff erosion by wave actions. Wave height ranges from 0.1 to 0.6 m with a wave period of 5–7 s during calm winter season, whereas these become 1.8–2.4 m and 12–14 s respectively during the rough summer session. During the period of cyclonic storms wave height goes above 2.4 m and wave period above 14 s.

1.7 Physiography and Configuration

The Sunderbans Delta Complex and its surroundings have a dynamic physiographical and geomorphological evolutionary history. The region, being a part of the Bengal Basin, represents coalesced multi-generation deltas that have prograded in phases during the positive interglacial eustatic sea level changes that occurred during the Plio-Pleistocene time towards the Bay of Bengal leaving behind distinctive multilevel delta surfaces, terraces, palaeochannels and palaeoshorelines migrating the successive coastlines towards sea. Between the main land and the sea, lies an immense archipelagic of innumerable islands-some are large, others small not bigger than sand bars, some persisted throughout the recorded history while others have just raised their heads into being.

Lying south of the Tropic of Cancer the total Sunderbans area of both Indian and Bangladesh part is bounded by Baleswar River on the east in Bangladesh and Hugli River on the West in India. Sunderbans is a kind of tropical humid forest. It may be considered as Bengalian rain forest biogeographically. Mangrove forests are occasionally called as 'Tidal forests' or 'Coastal wetlands'. The area of Sunderbans characterises a low flat alluvial plain covered with mangrove swamps and marshes, and is intersected by a large number of tidal rivers, estuaries, creeks and saltwater courses (Bagchi 1972). This network of drainage with dynamic flow patterns of tidal waters, the process of erosion and accretion of lands make the geomorphic set up of the area a complex one, when studied from north to south as well as from east to west.

The rivers of Sunderbans flow in an essentially southward direction with a funnel shaped opening in the Bay of Bengal. Both banks of these rivers and their distributaries exhibit phenomenon of slumping and terrace structures due to subsidence of marginal bank materials simultaneous with sediment compaction supporting mangroves in different stratigraphic levels. This subsidence might have relation to the neotectonic movements of the Bengal basin (Umitsu 1993, 1997).

Large portions of the river flood plains are inundated during rainy season but the flood plains are completely dried up during pre-monsoon times to be mud-cracked with surface salt encrustations. During dry seasons, the river water is confined in the channels.

The intertidal zone of the river flood plain generally is densely forested, whereas, the supratidal zone is occupied by dwarf mangroves. The regular inundation and exposure together with a freshwater discharge from the upland areas during monsoon is supposed to be the factors for the healthy growth of mangroves in the intertidal flats of the flood plain. The mangrove forests occupy both banks of the rivers and creeks (Fig. 1.1). But reclamation of these forest areas for inhabitation and cultivation is imposing a severe impact on the total ecosystem of the area.

1.8 Biogenous Coast

Tidal marshes in the upper intertidal flats occur in patches. The marshes are flooded twice daily during the high water spring to the high water neap tidal cycles with sediment-laden water. During this span of time the fine suspended particles of very fine sand, silt and clay are trapped into the Marsh surface and settle down gradually. Biological factors have an important role for the evolution in the coast and therefore, Sunderbans may be considered as biogenous coast. Bioturbation features are also very much significant for sediment trapping. The primary depositional behaviour of sediments is manipulated largely by different intensities of biogenic activities.

1.9 Pattern of Sedimentation

The Ganges – Brahmaputra – Meghna (GBM) river system while draining through the deltaic alluvial plain on their course to the Bay of Bengal carries an estimated annual sediment load of 2–4 billion tons (Coleman 1969). This sediment load is composed of sand, silt and clay. Sands settle at the confluence of rivers and the Bay of Bengal creating river mouth bar. Silts in suspension come back to the rivers again due to tidal force at mouth and ultimately settled down at riverbanks leading to the formation of mud flats. 'Swatch of No Ground' located at the abyssal plain of Bengal basin and tidal bore at most of the river mouths makes obstruction to the carried sediment load at sea. So, the sediments come back from the Bay of Bengal and depositional processes are continued along the course of the rivers and help in the formation of islands. In this way, tidal shoal, crescentic or linear point bars are formed. Vegetation consequently modifies the bar into the islands, which are isolated either by the creeks or tidal inlets. River Hugli changes its course several times due to tectonic rise and for this reason, the rivers of Sunderbans are no more connected at their heads with Hugli River and the fresh water supply is stopped there on.

1.10 Stratigraphy

The depositional behaviour with pattern of subsidence and newly accreted zone were studied across the tidal rivers and their distributaries in Sunderbans. The stratigraphic sections reveal differences in colour and grain size and different intensities of bioturbation in different lamination from bottom to top. There is in general coarsening of grain size towards the top in all stratigraphic sections. Hiatuses in sedimentation on stratigraphic breaks are marked depending on levels of oxidation, abrupt truncation of bioturbation zones on sudden appearance of bioturbation in a unit.

River sections reveal yellowish grey silt and clay, dark silt and clay and woody peat at different stratigraphic levels. The colour variations in the stratigraphic units truly reflect the differential rates of oxidation from top to bottom.

1.11 Geotectonic Settings

The Sunderbans delta complex extends geographically between the sea and plains of Bengal along the easternmost coast of India and transgressing Bangladesh having geo-genetic link to the southern fringe of Bengal basin (Fig. 1.3). The tectonic evolution of the greater Bengal Basin is fundamentally related to the collision pattern of the Indian plate with Burma and Tibetan (Eurasian) plates. The collision of these plates can be visualized in two different forms: (i) the north to northeasterly continent-continent collision of the Indian plate into the Tibetan plate, which is mainly expressed by thrusting, lateral displacements, and uplift associated with the development of eastern Himalayas; and (ii) the oblique subduction of the oceanic crust beneath the Burma plate resulting in the development of accretionary wedges, which together with thrusting and folding has subsequently uplifted the Indo-Burman Ranges (Sikder and Alam 2003).

The Bengal basin, being one of the world's widest, deepest and tectonically active basins, represents a classical asymmetric pericratonic basin that originated through different phases of the Tertiary Himalayan orogeny. Being bordered all around by tectonic fabric, the basin has a relatively stable shallow (1–8 km thick sediment) shelf part in the west and north west facing the Indian shield and a tectonically active southern and eastern foredeep part centred below the present Ganga-Brahmaputra river mouths. These two parts are separated by a hinge zone marked by high gravity and magnetic anomalies. The so-called Hinge Zone, a zone of presumed deep-seated normal faults in the basement complex, is conventionally thought of as representing the dividing line between Indian platform with full thickness of continental crust and the Bengal Fore deep. The position of the Hinge zone is usually shown to truncate against the Dauki Fault in the northeast. Another interpretation is that the Hinge zone passes somewhere through the Sylhet trough and probably continues towards the Halflong thrust at the northeastern corner of the Bengal Basin (Bilham and England 2001).

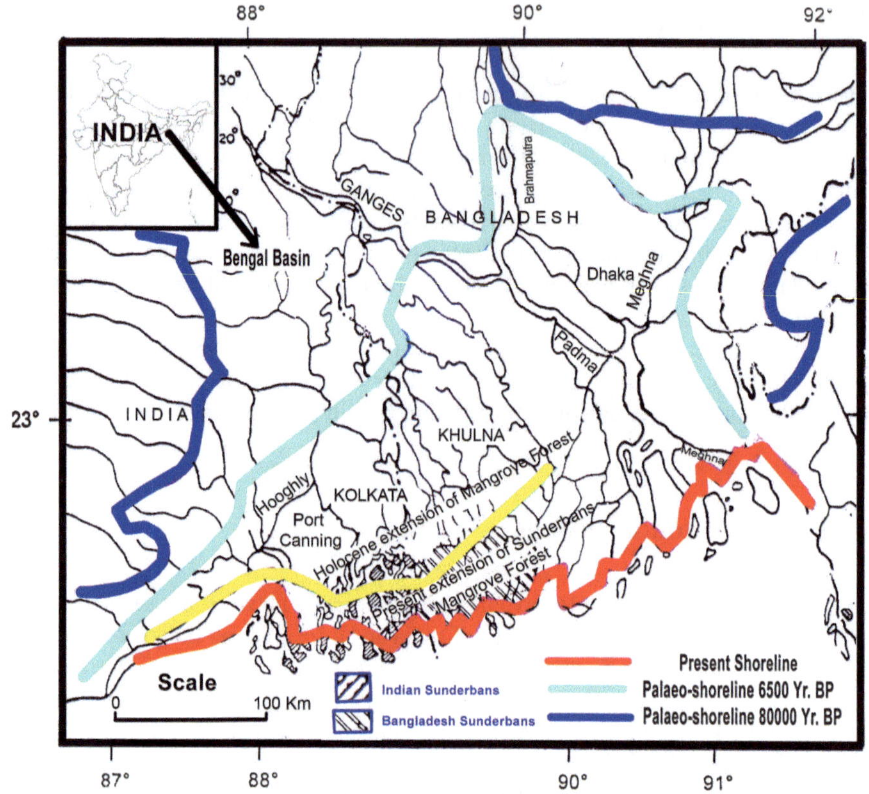

Fig. 1.3 Physiographic map of the Bengal Basin showing mangrove forest extension and successive migration of coastline through geological ages (Modified after Das 2006)

The Bengal Basin got filled up through Tertiary marine geosynclinals and shelf sedimentation (<16 km thick) followed by gradual progradation of the Quaternary Ganga-Brahmaputra delta fronts towards the southern sea producing the Bengal Delta complex, the mangrove vegetated Recent-Subrecent part of which is popularly known as the Sunderbans Delta Complex (Fig. 1.3). The Sunderbans Delta Complex is characterized by prolific growth of rich and diverse mangrove vegetation and forms an integral down drift coastal part of the Bengal Delta Complex that overlies huge thickness of Tertiary marine sediments of the actively subsiding Bengal Basin. The coastline of West Bengal belongs to the Amero trailing-edge coast (Inman and Nordstrom 1971; Davies 1972) whereas; the opposite continental coast is a collision coast. This plate-tectonics-related factor imposes an overriding control on the evolution of the coastline of West Bengal.

The present day drainage pattern of this coastal area results from two distinct factors noticeable to-date through the Tertiary history of Bengal basin (Biswas 1963; Sengupta 1966). These are: (i) a regional southerly slope of the Bengal basin due to the movement on the hinge zone at the edge of the shelf and (ii) increasing rate of southerly tilt of West Bengal part of the Bengal basin to a relatively greater rate of subsidence of the southwestern part of the hinge. This tectonic control of the basin is further accentuated due to the presence of numerous faults, which have intermittently active during the Quaternary (Morgan 1970). All the important river systems, of Sunderbans maintain a north to south trend and are actually surface manifestation of the subsurface southerly tilt of the Bengal basin. This southerly tilt, however, has been supposed to be more effective right from the late seventeenth century (Sengupta 1966, 1972).

The temporal as well as geographical distribution of mangrove plants are largely controlled by continental drift, dynamics of land vis- a-vis sea and environmental and physiographic changes over time since the advent of mangrove ecosystems on the earth in the tropical-subtropical coastal land masses after the break down of Gondwanaland,. The present Sunderbans mangroves have their wide spread ancient counterparts buried under deltaic sediment cover further inland. Being situated on the delta of the Ganga River, the accretion erosion behaviours of the rivers are primarily controlled by the regional geology of delta formation involving both fluvial and coastal processes. The deltaic zone represents a downwarped basin flanked by successively occurring older rocks from Tertiary and older sediments to younger Pleistocene Terraces into which the major river valleys have been incised (Morgan 1970). High macrotidal amplitude (4.6–5.5 spring tide) in the coast results in strong tidal currents leading to a network of deeply scoured (>30 m) tidal rivers and inlets and formation of an overlapping tidal plain across the intermittently subsiding Bengal delta (Morgan 1970).

1.11.1 Bengal Basin

The Bengal Basin is well known for the development of a thick (>22 km) Early Cretaceous – Holocene sedimentary succession (Curray 1991). The Bengal Basin in the northeastern part of Indian subcontinent, between the Indian shield to the west and north, and the Indo-Burman Ranges to the east, covers Bangladesh, part of West Bengal and Tripura states of India and the Bay of Bengal. The Bengal basin is an asymmetric depression that deepens towards the adjacent Indo-Burmese orogenic belt in the east. A vast flat alluvial plain, encompassing West Bengal, Assam, Tripura of India and adjacent Bangladesh forms the surface of the Bengal Basin. It is an asymmetric depression on a crest dipping to the east under the tectonic load of the Indo-Burmese thrust belt located between the eastern edge of the main Indian craton and Indo-Burmese orogen and south of the Shillong plateau. The two parts of the basin are separated by a hinge zone marked by high gravity and magnetic anomalies (Sengupta 1966).

The Bengal basin primarily refers to the western, relatively shallower part of the basin which lies inside the Indian Territory. The Bengal basin encompasses an area of 56,000 km^2 on-land and 33,700 km^2 in the offshore upto 200 m isobaths to the south and it extends further to the east in Bangladesh. The Bengal basin extends from the basin margin fault east of the Chotonagpur plateau in the west, Shillong plateau in the north bounded by Dauki strike-slip fault, and Indo-Burmese thrust belt in the east. The total sediment fill in the basin during Cenozoic, even for the limited area of the basin which lies inside the Indian territory, is over 315,000 km^3. More than half of the Cenozoic sedimentary volume was deposited during the Plio-Pleistocene period leading to the southward growth and development of the modern Bengal delta. The present asymmetric geometry of the Bengal basin developed primarily due to flexure of the lithosphere under the bounding thrust belt and the associated tectonic load in the post-collision time.

The Bengal Basin in the northeastern Indian subcontinent evolved from a passive continental margin (Pre- Oligocene) to a remnant ocean basin (beginning in Miocene) comprising three geo-tectonic provinces: (i) passive to extensional cratonic margin in the west, the stable shelf; (ii) the central deep basin or remnant ocean basin; (iii) the collision and sub-duction-related orogen in the east, the Chittagong-Tripura fold belt. These geo-provinces have been related to a regional plate tectonic scenario, especially the collision pattern of the Indian plate with the Burma and Tibetan (Eurasian) plates.

1.11.2 Bengal Delta

The Bengal delta is one of the largest tide-dominated river deltas in the world. The terrigenous sediments transported to the southern margin of the delta by the Ganges-Brahmaputra River systems are being constantly reworked and redistributed by the strong tidal currents from the Bay of Bengal. The huge Bengal delta is made up of the combinations of two processes which may have operated through the Plio-Pleistocene period-(i) discharge of great volumes of clastic sediments through the Bengal trough by the Ganga-Brahmaputra drainage system due to denudation of the rapidly rising Himalayan chain in the north and (ii) a high rate of Bengal basin-floor subsidence nearly along a N-S axis causing selective deposition of coarser clastics in the Bengal basin adjacent to the Indo-Burmese thrust front. The latter allowed onward transport of predominantly finer terrigenous clastics to a large part of the Bengal delta.

1.11.3 Deltaic Flood Plain

Generally both the higher-elevation and fresh water upper delta plain and brackish water lower delta plain form sub-aerial facies as a result of delta progradation. Fresh water marshes, poorly drained swamps and lakes altogether lead to the formation of upper deltaic plain of the Ganga-Brahmaputra. This delta with low flat alluvial plain is an example of low land flood plain. Sunderbans with Holocene alluvial stratigraphy of an average 40 m in thickness belongs to the low land flood plain.

The upper delta plain has been divided into 17–20 distinct flood plains (Allison 1998; Alam et al. 2003). Flooding characteristics of rivers, flood plain elevation, local subsidence, sediment compaction, tectonic uplift and subsidence and construction of embankments control the receiving of sediments, its characteristics and quantity (Allison 1998). Sediments in the flood plains is dominantly silty with a tendency of decreasing of grain size away from the channels (Coleman 1969). Low land flood plain is characterized by the wetlands permanently filled in with the waters which are formed probably as a result of tectonic and compaction-induced subsidence (Allison 1998).

1.11.4 Lower Delta Plain

The present situation of the Sunderbans is in the Hugli-Matla estuary of the western part of the lower delta plain (tidal) of the Ganga-Brahmaputra discharge. The maximum sea level transgression occurred at about 6,500 BP in terms of all available evidences when the shoreline was 100–300 km inland of the present shoreline. Banerjee and Sen (1988) in their investigation of the sub-surface Holocene sediments around Calcutta came across floral and faunal assemblages indicative of the tidal mangrove forest flourishing further north of the present Sunderbans about 6,000–7,000 years BP (Fig. 1.3). Subsequent to this time, basin infilling and shoreline progradation accounts for about 30 % of the modern delta growth by the Ganga-Brahmaputra rivers where the paleo-shoreline (Fig. 1.3) follows approximately the present 3 m elevation contour (Allison 1998).

1.12 Geomorphic Environments

Estuarine Sunderbans is a complex net work of tidal channels, extensive tidal flats, mangrove swamps, low salt marshes and islands. Process of erosion and deposition, morphologic variability and sediment transport determines the geomorphology of this estuarine river system (Ginsberg and Perillo 2004). A complex network of geomorphic environments characterises Sunderbans. The important morphotypes of the Sunderbans are beaches and inlets, creeks and estuaries, mangrove swamps, mudflats, coastal dunes and sand flats. Both sandy beaches and mudflats occur at the sea face depending on their locations on high and low energy zones respectively. Tidal rivers/inlets are generally muddy systems barring some sandy flats that occur in the mid channel bars and sandy swash bars at the mouths of rivers/inlets. The creeks are absolutely muddy systems. Mangrove swamps occur on the intertidal mudflats of estuaries, creeks and inlets. Narrow marshes occur on the upper intertidal to the supratidal zones in a sporadic manner. Broad geomorphic environments can be recognised in Sunderbans like coastal areas, estuarine tidal river systems, tidal creek systems with mudflats, sandy delta beaches etc. The important geomorphic features are natural levees, point bars, mid channel bars, sand flats, coastal dunes, rivers, creeks and estuaries, beaches and inlets, mangrove river mouth and mud flats. The 130 km long coastline is changing showing indented type of longitudinal bars due to cliff erosion.

1.12.1 Coastal Zone

The meso-macrotidal coastline of West Bengal is comprised of tidal plains of a large number of estuaries and tidal inlets of which the tidal flats of Hugli and Matla estuaries are the most important. Tidal flooding, river fluxes, waves, long-shore currents, sea-level changes, regular episodic and non-episodic events like cyclones and storms are dominant physical processes that continually modify the coast in a short and long term basis. Long-term tectonic and geological processes (Biswas 1963; Sengupta 1972; Allison 1998; Kuehl et al. 1997; Bhattacharya 1999) further accentuate morphodynamic changes in the coastline. All the rivers and inlets widen into funnels as they approach the Bay of Bengal (Fig. 1.1). This indented estuarine feature of this deltaic coastline has marked the dominant role of tides of this morphological evolution of this coast and may be designated as "estuarine delta coast" (Galloway 1975).

1.12.2 Sandy Beaches

The foreshore sandy beaches of Sunderbans (e.g. Gangasagar of Sagar island, Kalas of Bulchery Island and Fraserganj of Namkhana) range in width from 100 to 350 m with a seaward gradient ranging from 1 in 50 to 1 in 90 (1.4–0.7°). In areas of ridge and runnel slopes, the gradient is as steep as 1 in 9(6°). The siliciclastic beach material is dominantly quartzo-feldspathic (95 %) with about 5 % accessory biotite, muscovite, hornblende and ilmenite. These beaches are made up of fine to very fine sand with good to very good sorting of sediments (Bhattacharya 1993). There is in general a negative correlation between grain size, wave energy or beach slope. Following Flemming and Fricke (1983) it has been estimated that for measured values of foreshore beach slopes ranging from 1:50 to 1:90, mean size ranges from 2.9 to 3.1 phi. The foreshore beaches show a number of surficial physical structures that include backwash ripples, rhomboid ripples, backwash marks, swash marks, current crescents and flat topped ripples. The creeks within foreshore contain unidirectional current ripples of both linguoid and lunate types. Interference ripples (ladder-back type) of wave and current origin abound in the lower foreshore zone (Komar 1976). Beach laminations with alternations of light and dark colours and gentle dips (1–4°) towards the seaward direction are the most common stratification types. According to the morphodynamic scheme of beach classification (Wright and

Short 1982) most of the beaches belong to the dissipative dynamic state having gentle but rhythmic beach profiles (Das and Bhattacharya 2002).

1.12.3 Islands

There are 102 islands of which 48 islands are with forestry and rest 54 islands are reclaimed and human habitated zone with agricultural land. In estuarine region of Sunderbans tidal river water courses enclose a large number of islands of various shapes and sizes modified after river mouth bar. Most of these islands are covered by marshy or swampy vegetation of dwarf mangroves at the centre and luxuriant mangroves at the girdle. These island of Sunderbans are characterised by the low forest and scrub wood jungles. The main estuarine islands situated at the mouths of the rivers from West to East are (i) Nayachar, Ghoramara, Jambu Island and Sagar island between Muriganga or Bartala and Hugli river (ii) Mohisani island between Chenair (Pitt's creek) and Muriganga (iii) Namkhana, Fraserganj and Bakkhali between Saptamukhi and Chenair (Pitt's creek) (iv) Henry's and Frederick Island between Saptamukhi and Bakkhali river (v) Lothian and Prentice Islands in the Saptamukhi (vi) Dhanchi Island between Thakuran and Jagadal river (vii) Bulcherry Island between Matla and Thakuran river (viii) Halliday Island in the Matla (ix) Dalhousie Island between Matla and Guasuba and (x) Bangaduni Island in the Bangaduni River. Three chief mechanisms of island formations are noticed in the deltaic Sunderbans – (i) depositional islands originated from settling down of particles on the river bed to form the mid channel bars and the Channel Islands. Being formed primarily under the influence of flood currents these flood-tidal bars have tendency to move upstream. The river mouth island (bar) formed by the influence of ebb currents, however, has a tendency to move seaward (ii) mid channel bar evolved by disconnection of the bank ward edge of the bar is occupied by active or abandoned spill channels and (iii) islands formed by mainland cut off during bifurcation of channels resulting from shallowing of river beds (Fig. 1.4). This is associated with intermittent subsidence and sedimentation of the Bengal delta.

1.13 Sunderbans: Its Uniqueness

(i) Sunderbans of India along with Bangladesh is the single largest mangrove forest of the world.
(ii) Sunderbans forest acts as a sink for the mega-metropolitan pollutants.
(iii) Sunderbans is the only mangrove forest that supports the largest tiger population in the country.
(iv) Tidal amplitude and fluctuations is very high in the river mouths of Sunderbans.

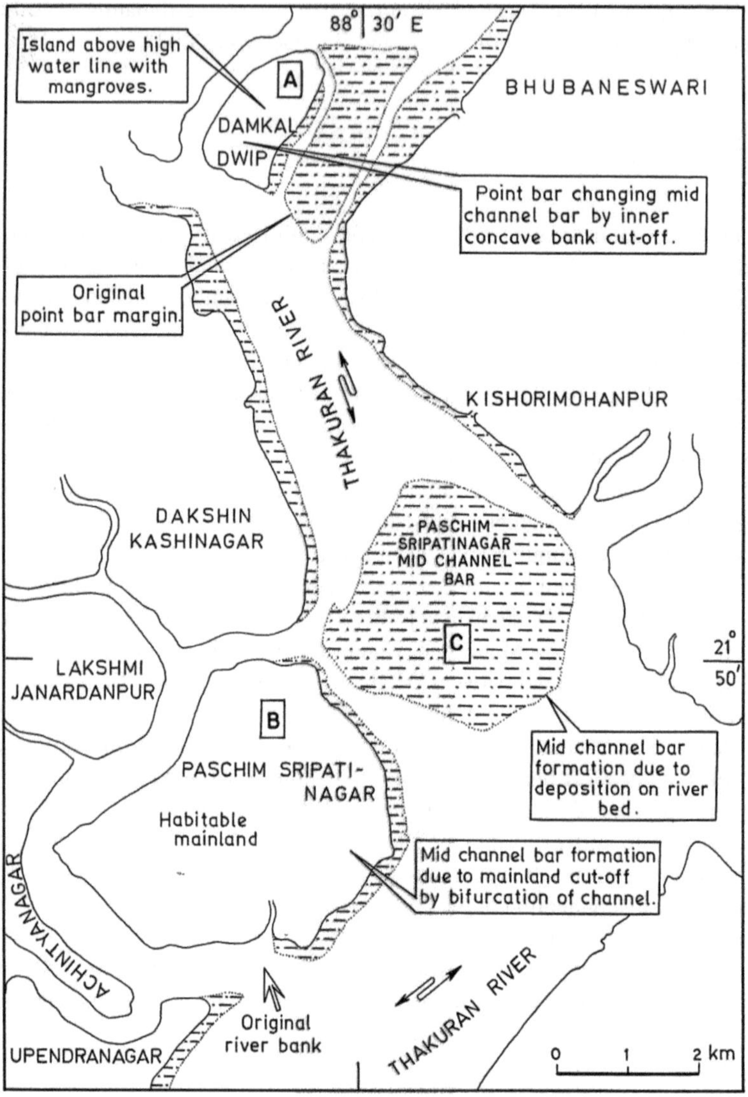

Fig. 1.4 Different mechanisms of island formations in the river of Sunderbans

(v) The mangroves of Sunderbans provide protection to island habitations against the fury of cyclones.
(vi) Sunderbans is the home of river terrapin which was once believed to be extinct.
(vii) It is the only mangrove forest having largest floral and faunal diversity in the world.
(viii) Sunderbans is the home of several endangered, threatened and endemic floral and faunal species.

1.14 Summary

The indentation of West Bengal coast is primarily because of the outfall of some important rivers and inlets. From the Indo-Bangladesh border in the east to farther west these include Ichhamati-Raimangal, Haribhanga, Gosaba, Bangaduni, Matla, Thakuran, Jagadal, Saptamukhi, Edward's Creek, Muriganga and Hugli. A number of geomorphological environments are associated with these fluvio-tidal and tidal systems and the present day configuration of this region is a reflection of the late Holocene balance between the fresh and salt-water courses in this delta under the hostile settings of both natural factors (tilting of delta, fluvial regime, state of coastal processes, sea level rise) and anthropogenic influences. The present paper deals with the role of several dynamic geomorphic processes and products of erosion and deposition to control the changing configuration of the topography and shoreline of the deltaic Sunderbans.

In a beach face there is a general trend of successive variations of ripple type along with other associated surface structures. The lower limit of the foreshore zone characterized by oscillatory flow is seen to be replaced by symmetrical flow more and more from mid foreshore to upper foreshore region. This reveals an increasing flow power as registered and reflected from the changing pattern of bed forms. Moreover, these bed forms indicate the flow regime change (Harms and Fahenstock 1965) in the beach face which starts from lower part of the lower flow regime in the lower foreshore zone through the upper part of the lower flow regime in the mid foreshore zone to the transitional to upper flow regime in the upper foreshore zone.

The meso-macrotidal regime of the river and its meandering course have further been controlled the development of sand bodies like linear bars, point bars, mid-channel bars and mouth bars.

Subsidence of the bank with simultaneous sediment filling is a common feature along the river course. The process of deposition in this tidal channel is chiefly controlled by high rate of sediment trapping in areas of geomorphic highs with network of mangrove vegetation. The process of bank erosion, however, is governed by the helicoidal flow couple with the sinuous ebb and flood currents. Near the mouth of the river, the process of erosion is further promoted by a gradual eating up of the bank materials in a piecemeal manner. The presence of these scenarios within the marginal waters of the rivers as well in the islands bears evidences of environmental morpho-dynamics of the estuarine semi-active delta, Sunderbans.

References

Alam M, Curray JR, Alam MM, Chowdhury MLR, Gani MR (2003) An overview of the sedimentary geology of the Bengal basin in relation to the regional tectonic framework and basin-fill history. Sediment Geol 155:179–208

Allison MA (1998) Geological framework and environmental status of the Ganges- Brahmaputra delta. J Coast Res 14(3):826–836

Anonymous (2003) A handbook on Sunderbans Biosphere Reserve. Ed. Director, SBR, Kolkata
Bagchi KG (ed) (1972) The Bhagirathi – Hooghly Basin. Calcutta University Publ., Calcutta
Banerjee M, Sen PK (1988) Paleobiology and environment of deposition of Holocene sediments of the Bengal Basin, India. In: Palaeoenvironment of East Asia from the mid- tertiary: proceedings of the second conference. Centre of Asian Studies, University of Hong Kong, Hong Kong, pp 703–731
Bhattacharya A (1993) Backwash and swash oriented current crescents: indicators of beach slope, current direction and environment. Sediment Geol 84:139–148
Bhattacharya A (1999) Embankments and their ecological impacts: a case study from the tropical low-lying coastal plains of the deltaic Sunderbans, India. In: Vollmer M, Grann H (eds) Large-scale constructions in coastal environments. Springer, Berlin, pp 171–180
Bilham R, England P (2001) Plateau 'pop up' in the great 1897 Assam earthquake. Nature 410:806–809
Biswas B (1963) Results of exploration for petroleum in the Western part of Bengal basin, India. Proc 2nd Symp Dev Ptr Res ECAFE Min Res Div Ser 18(1):241–250
Coleman JM (1969) Brahmaputra River: channel process on sedimentation. Sediment Petrol 3((2/3) Spl Issue):129–139
Curray JR (1991) Geological history of the Bengal geosynclines. J Assoc Explor Geophys XII:209–219
Das GK (2006) Sunderbans-environment and ecosystem. Levant, Kolkata
Das GK (2009) Beach processes of Tidal Islands of Hugli Estuary, West Bengal. Geogr Rev India 71(3):240–248
Das GK, Bhattacharya A (2002) Dynamic geomorphic environment of Indian Sunderbans. In: Basu SR (ed) Changing environmental scenario of the Indian sub-continent. ACB Publication, Kolkata, pp 284–298
Davies JL (1964) A morphogenetic approach to world shoreline. Zeits Fur Geomorph 8:127–142
Davies JL (1972) Geographical variation in coastal development. Oliver & Boyd, Edinburg
De Boer PL, Oost AP, Visser MJ (1989) The diurnal inequality of the tide as parameter for recognizing tidal influences. J Sediment Petrol 59:912–921
Flemming BW, Fricke AH (1983) Beach and nearshore habitat as a function of internal geometry, primary sedimentary structures and grain size. In: McLachlann A, Erasmus T (eds) Sandy beaches as ecosystems. Dr. W. Junk Publ., The Hague
Galloway WE (1975) Process framework for describing the morphologic and stratigraphic evolution of deltaic depositional systems. In: Broussard ML (ed) Deltas, models for exploration. Houston Geological Society, Houston, pp 87–99
Ginsberg SS, Perillo GME (2004) Characteristics of tidal channels in a mesotidal estuary of Argentina. J Coast Res 20(2):489–497
Harms JC, Fahenstock RK (1965) Stratification, Bed forms and flow phenomena with an example of Rio grade in Primary Sedimentary Structures and their hydrodynamic interpretation, S.E.M.M. Spec. publ. 12., pp 84–115
Inman DL, Nordstrom CE (1971) On the tectonic and morphologic classification of coasts. J Geol 79:1–21
IPCC (2007) Climate change-2007. Impacts, adaptations and vulnerability. Contribution of working group II to the fourth assessment report of the Intergovernmental Panel on Climate Change. Cambridge University Press, Cambridge
Komar PD (1976) Beach process and sedimentation. Prentice Hall Inc, Englewood Cliffs
Kuehl SL, Levy BM, Moore WS, Allison MA (1997) Subaqueous delta of the Ganges- Brahmaputra river system. Mar Geol 144:81–96
Morgan JP (ed) (1970) Deltaic sedimentation: modern and ancient. Special Publication Society of Economic Paleontologists and Mineralogists 15, Society of Economic Paleontologists and Mineralogists, Tulsa, 312p
Sengupta S (1966) Geological and geophysical studies in the western part of Bengal basin, India. Am Assoc Petrol Geol Bull 50:1001–1017

Sengupta S (1972) Geological framework of the Bhagirathi-Hugli Basin. The Bhagirathi-Hugli Basin proceedings of the inter disciplinary symposium. Calcutta University Publications, Calcutta, pp 3–8

Sharma AP (2010) Coastal zone vegetation in India with reference to mangroves and need for their conservation. In Sunderbans: issues and threats. CIFRI (ICAR), Kolkata, pp 9–40

Sikdar AM, Alam MM (2003) 2-D modeling of the anticlinical structures and structural development of the eastern fold belt of the Bengal basin, Bangladesh. Sediment Geol 155:179–208

Umitsu M (1993) Late Quaternary sedimentary environments and landforms in the Ganges delta. Sediment Geol 83:164–178

Umitsu M (1997) Landforms and floods in the Ganges delta and coastal lowland of Bangladesh. Mar Geol 20(1):76–87

Vannucci M (1989) The mangroves and us; a synthesis of insights. Indian Association for the Advancement of Science, New Delhi

Wright ID, Short AD (1982) Morphodynamics of beaches and surfaces in Australia. In: Komar PD (ed) Hand book Coastal processes and erosions. CRC Press, Washington, DC

Chapter 2
River Systems and Tidal Courses

Abstract The Sunderbans are drained by numerous interconnected rivulets, tidal creeks or inlets and estuaries. The tides and waves in these water bodies coupled with other geological and tectonic processes are responsible for significant morpho-dynamic changes in the macrotidal coastline. Extensive blanket of flood plain deposit, low ridges of natural levees, channel bars associated with various environmentally significant sedimentary structures characterise the lower estuarine zone. The other conspicuous landforms are mudflats, back swamps, salt marshes, tidal creeks with mangrove, coastal dunes, beaches, tidal flats etc. Construction of embankment for protection of erosion in Sunderbans is found to have an adverse impact on the fragile ecology of the Sunderbans.

Keywords Sunderbans • Tidal rivers • Channels • Estuary • Bedforms • Point bars • Levees • Mudflats

The network of numerous tidal water courses in the Sunderbans are connected with each other by an intricate mesh of channels, so that the whole tract is a tangled networks of creeks, inlets, rivers and estuaries. The mouths of the tidal rivers at the confluence of the Bay of Bengal are wide funnel-shaped and those tidal mouths are called estuaries. The main estuaries in the Indian Sunderbans are with the rivers Muriganga or Bartala in the west, Saptamukhi, Thakuran, Matla, Bangaduni, Guasuba and Haribhanga in the east (Sharma 2010).

Estuary is a partially enclosed coastal body of water is significantly diluted by fresh water from land runoff. Estuaries are originated due to sea level rise. Sea level has risen approximately 120 m over the past 18,000 years due to melting of glaciers. All the estuaries of Sunderbans are characteristically coastal plain estuary. A coastal plain estuary forms as rising sea level causes the oceans to invade existing river valleys.

Tidal courses and river channels are considered as the main features of the estuarine environment because they contain areas of important ecological values, which are extremely sensitive to changes caused either by natural factors or anthropogenic interferences (Ginsberg and Perillo 2004). Thirty-one rivers forming a network of intricate minor salt-water courses intersect the Indian Sunderbans. These rivers and tidal inlets serve dual purposes of freshwater discharge and to and

G.K. Das, *Estuarine Morphodynamics of the Sunderbans*,
Coastal Research Library 11, DOI 10.1007/978-3-319-11343-2_2

fro movement of saline water wedges arising from the sea face. The coastline of West Bengal is comprised of the tidal plains of a large number of estuaries and tidal inlets of which the Ganges Brahmaputra tidal plain is the most important. Tidal flooding, river fluxes, waves, sea level change, regular episodic and non episodic events like cyclones and storms are the dominant physical processes that continually modify this coast in a short and long term basis.

2.1 River Networks

So-called tidal rivers, 31 in number of Indian Sunderbans are basically large tidal creeks in this low-lying coastal deltaic plain. They have no perennial fresh water sources. Most of these tidal rivers are the result of tidal incursion and retreat. During the process of to-and-fro movement of tidal water on a gently sloping plain these rivers are joined by many other creeks of lesser and lesser magnitude variously called locally as khal, gang etc. The orientation, geomorphic setting and materials of the river deposits are resultant products of the complex interaction of several factors like tectonic frame work of the Bengal basin, geological, climatological, physical, chemical and biological processes. The geographic location of the rivers also acts as important parameters as all other factors are subjected to change depending on its variability. The morphodynamic changes of the coastline are supposed to be the outcome of (i) long-term events like tectonic and geological process and (ii) Short-term events like tides and waves. Both these events act jointly in this highly dynamic macrotidal coastline.

The river mouth widens into funnel as the river approach the Bay of Bengal. This riverine feature characterises this coast of West Bengal to be designated as a deltaic estuarine coast. There are in all 31 major tidal rivers and estuaries and many minor inlets and creeks, which have given rise to a much indented nature of this coast. From the Indo-Bangladesh border in the east to further west, the important rivers and creeks are Roymangal, Haribhanga, Jhilla, Guasuba, Bangaduni, Matla, Thakuran, Jagadal, Saptamukhi, Mridangabhanga, Gobadia, Edward's creek, Chenair, Muriganga or Bartala and Hooghly (Figs. 2.1 and 2.2).

An asymmetric distribution of flow velocity over transverse profiles of these meandering rivers facilitates erosion on the outer concave side and deposition on the inner concave side of the meander bends. Point bars, thus, are the products of lateral accretion and are resultant from the combination of helicoidal and sinuous tidal flows (flood and ebb) of the rivers. This flow complex also results in the indentation of river banks in many cases.

A large number of meandering creeks emerge at right angles to the periphery. These creeks are of variable length and width. Down slope flow of water through these creeks during receding water creates collapsing of the creek margin. The surface of the mid channel bar emerged on the river bed is characterised by flood and ebb dominated sedimentary structures like ripple marks, megaripple marks of

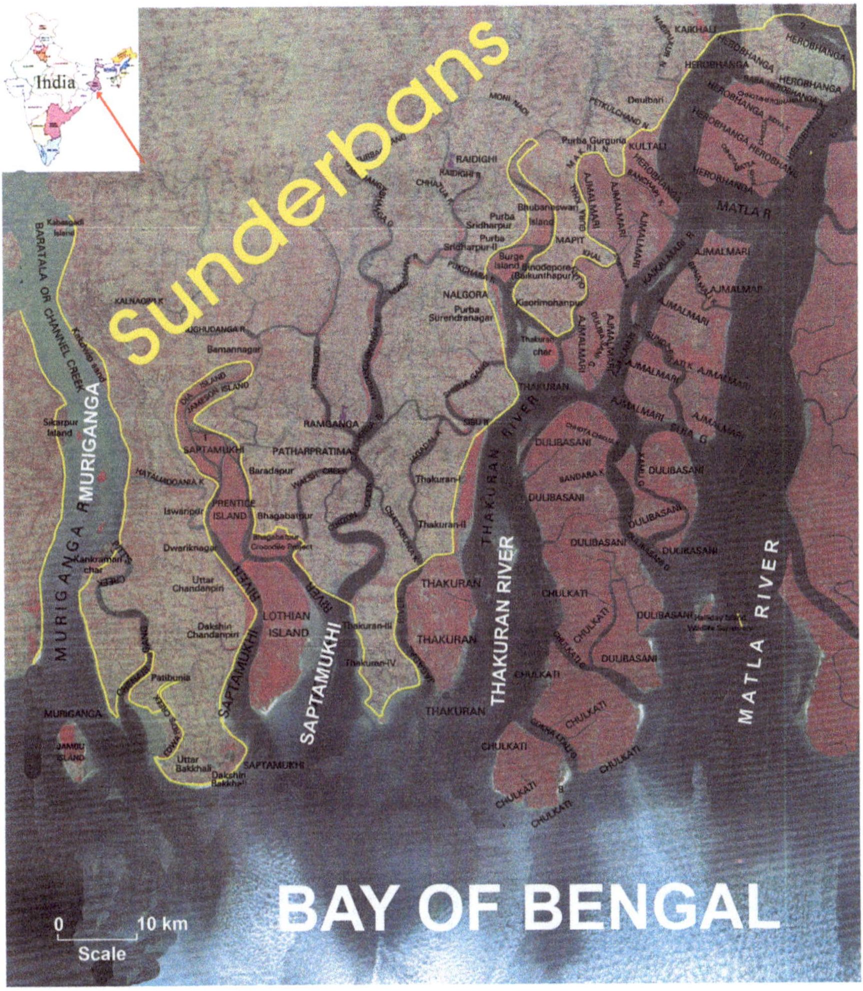

Fig. 2.1 Map of western part of Sunderbans showing the estuarine rivers and other tidal water courses

various kind and magnitudes. All these bedforms suggest that from lower part to the upper part of the lower flow regime of the rivers (Simon and Richardson 1962) are characteristics of this area.

2.1.1 Ichamati

The river Jamuna that is joined by the Ichamati near Tibi of Nadia district directly feeds Ichamati. Ichamati serves as the international borderline between India

Fig. 2.2 Intricate tidal river networks in the eastern part of Sunderbans

and Bangladesh and again it makes the boundary between Nadia and North 24 Parganas district just before Bagda P.S. At Hingalgunj it enters into the Dampier – Hodges imaginary boundary line demarcating Sunderbans and it gives numerous distributaries namely the Raimangal, Bidya, Jhilla, Kalindi, which are wide funnel shaped estuaries in the Sunderbans.

Of late waters of Ichamati overflow after a short spell of rain resulting flood during September–October every year due to heavy siltation, setting up of fisheries by blocking the river with wooden rafts, brick field and houses built up on the riverbed, growth of water hyacinth blocking river flow, local markets coming up etc. No full-fledged de-siltation drive has been undertaken. So the threat of flooding has increased considerably along the large stretch of Ichamati.

2.1.2 Kalindi- Raimangal -Haribhanga

The stretch of Ichamati is bifurcated into two separate channels at Hingalganj the Western smaller channel known as Sahebkhali and the main flow at the eastern side is known as Kalindi. Bidyadhari and Sahebkhali River are joined at Sandeshkhali Block 2 and form the course of Raimangal River. Raimangal flows southward and then merges with the river Kalindi. The combined flow of Raimangal meets the Bay of Bengal in the south as Haribhanga River which serves as the international boundary between Bangladesh and India in the south-east.

2.1.3 Bidyadhari

Like many other drainage systems of North and South 24 parganas, West Bengal the drainage systems of the Bidyadhari river is under great threat due to stresses arising from both natural and anthropogenic sources. Bidyadhari River occurs in at least four different areas. Of these, the main stream of Bidyadhari, this has ultimately bifurcated into Raimangal and Jhilla rivers before it debauches into the Bay of Bengal. The upstream part the river is actually a combination of north south flowing Nawai, Sunti and Nona gang of which the last one was later renamed as Bidyadhari. Before 100 years all these drainage systems were fed by the upland freshwater discharge from the Hooghly and Padma rivers. These situations for this river system however, changed with time when all these river systems were cut-off from the headwater discharge and are now only active during the rainy season being fed by the precipitation in the catchment areas.

Further, west of this mainstream, the Bidyadhari River occurring in the upstream of Matla River takes a southeasterly trend to join the Karatoya and Atharabanka rivers. This Bidyadhari drainage system has completely died out during the last century and very little of it can hardly be identified as a drainage channel. The Matla River that flowed south to the sea was also navigable by steamer upto Canning Port that time. A canal named Bidyadhari khal trending northeast –southwest join the main and the second Bidyadhari channel. The third Bidyadhari drainage east of the main stream is also known as Kurnal River which is almost a non-existing system at the present time.

The main Bidyadhari river system has taken four different names from Haroa as Haroa River, Chaumuha River, Malancha River and Kalagachia River from north to south till it has come down to Dhamakhali for a total stretch of 120 km.

The main Bidyadhari River is tidal upto a length of 95 km and the ebb and flood tidal currents play semidiurnally within the channel systems. Barring the precipitated fresh water from the catchment area there is no fresh water discharge. On the contrary, the central part of the Bidyadhari River around Ghusighata serves as an outfall channel for the sewage of the city of Kolkata mainly through Bagjola and Bhangarkata khal.

Other human activities involving inland fisheries, brick making, tannery industry etc. further worsen the condition of the river system bringing about a total degradation in the ecological condition of the drainage basin. The rate of siltation on the river bed has increased to an alarming extent because of transportation of material due to land reclamation, cultivation, dumping of sewage sludge and fisheries wastes.

The dwindling patterns of mangrove vegetation barring certain pollution tolerant species e.g. *Sonneriatia* sp. & *Avicennia* sp. and some macrobenthic animals like crab *Uca* sp. and polychaetes worms *Mastrobranchus indicus* and *Dendronereis heteropoda* have become most conspicuous for the last two decades. The soft bottom polychaetes particularly, act as opportunistic species in this area eliminating the other polychaete species.

Thus, the Bidyadhari river catchment is losing its life rapidly. Along with a stupendous change in its ecological behaviour, its capacity of holding the sewage is also coming to a halt. Some remedial measures have also been suggested for reviving its morphodynamic and ecological character of the river.

2.1.4 Piyali

Piyali branches off the river Bidyadhari near Pratapnagar in the South 24 Parganas district and flows about 45 km to the south and southwest till it joins the river Matla about 37 km below Port Canning. A pumping station is set up near Uttarbhag Ghat with the financial assistance of Govt. of West Bengal, which helps in drainage of land run off water from the area of Garia, Sonarpur, Subhasgram, and Baruipur to make these areas habitable. Of late, the large dam constructed near Ambikanagar for storing rain water for irrigating agricultural lands has blocked the southern part of the course of Piyali River. The closure of river mouth prevents navigation; migration of fish and consequently the riverbed is filled up with the water hyacinth. At present Piyali is a dying river.

2.1.5 *Matla*

Apart from Hugli River, Matla seems to be the next wide estuary in the Sunderbans region. The combined flow of Karati in the north and Amjhara in the east of the Bidyadhari forms the course of Matla near Port Canning (Fig. 2.1). Matla is the local name of Port Canning. The river is named after Matla region. The Matla meets the Bay of Bengal about 120 km towards south direction.

Network of channels connected with Matla

West bank	East bank
1. Piyali R. and Kultala gang forming inland drainage basin.	1. Hogal at Sonakhali named later Pathankhali river at Pathankhali
2. Nabipukur creek – a channel	2. Bidya Khal joins Sialfeli creek
3. Bainchapi creek takes the name of Gura Khal later.	3. Lotaman and Jhara creek- narrow channels.
4. Kankalmari Ajmalmari suia gang are the major tidal creeks.	4. Small and large Herobhanga are tidal creeks.
5. Kamli and Dulibhasani gang named from Suia gang	5. Another creeks Chota Matla and Chota Bidya
6. Chulkati, Parsemari, and Jhimara gang originated from Dulibhasani gang	6. Bidya river joins Matla
7. Gokultali gang connects with the Thakuran river	7. Numerous tidal inlets Netidhopani, Storkhali, Suryamukhi, Havatidunia, Haldi and Gobdi Creeks.
	8. Mayadwip river – a major tidal creek.
	9. Choubanka, Lakshmi banka, Chhaimari, Chota and Bara Arhbhanga Khal are at the extreme south.

2.1.6 *Bidya*

Combined flow of narrow Huta Khal and large Durgamandal tidal creek takes the names of river Bidya in the north. Bidya flows about 55 km southward and crosses two major blocks namely Basanti and Gosaba of South 24 Parganas. Bidya merges with the river Matla near Herobhanga Reserve Forest at Jharkhali.

Channels and creeks connected with Bidya River

West bank	East bank
1. Hatakhali – named as Pathankhali creek later	1. Melmel Khal and Huta Khal are narrow creeks.
2. Radhanagar and Durgamandal creek-major creeks in the south.	2. Gumdi and Durga Doania tidal channels near Sajnekhali Sanctuary.
3. Moukhali nala and Kartal gang are connected	3. Nabanki creek -a major creek
4. Dundul khali creek a narrow channel	4. Netidhopani- a major tidal creek.
5. Kayer Creek and Mocumberiya creeks are joined.	

2.1.7 *Guasuba*

Jhilla River is bifurcated into two separate channels, the eastern channel is the Haribhanga R. and the western channel is Guasuba River. The Guasuba river flows entirely through the Sunderbans Reserved Forest and meets the Bay of Bengal after flowing south ward for about 30 km.

2.1.8 *Bangaduni*

Bangaduni River is formed with the combined flows of Haldi creek and Mayadwip River. It flows only 6 km southward and meets the Bay of Bengal. Bangaduni Island is situated in the Bangaduni River (Fig. 2.2).

2.1.9 *Thakuran*

The Thakuran River is an important drainage basin over the basement of alluvial and tidal sediments and has a length of 80 km from north to south. From a very small width of less than 1 km at its northern uppermost part, the funneled mouth of the river attains a width of about 8 km before meeting the Bay of Bengal in the south (Fig. 2.1). The river is entirely tidal and belongs to a macro-mesotidal regime from its mouth (5.5–4.6 m spring tide) further upstream (3.2–2.8 m spring tide). The basin occupies approximately 900 km^2 areas and is characterised by a number of small islands separated from one another by numerous tidal waterways. It belongs to southerly sloping lowland and is flanked by fringing mangrove swamps all along the banks covering the interchannel areas.

The Thakuran River is not directly fed by any freshwater from its landward side like the Ganges. Rainwater and run-off water from the catchment are the only source of freshwater in it. Thus, the decrease in salinity of the Thakuran River water during monsoon reflects the effect of high monsoonal precipitation and subsequent run-off of rainwater into the river basin. The middle stretch of the river is bifurcated into two separate channels, the western smaller channel known as Jagadal gang (river) and the main flow in the east as the Thakuran River. Both these bifurcated channels have a more or less parallel north-south trend. The Chirapat or Pakhirala creek connects the Jagadal gang at its upper end to the Thakuran River. Beyond this point of bifurcation further north, the Thakuran gradually shows lesser and lesser width and ultimately connects the Matla River through the Nobipukur-Piali, a distributary of Matla. At this point there is an outfall of another inland channel called Nimania Khal (creek) from the northwest.

Network of channels and creeks connected with the Thakuran River from north to south

West bank	East bank
1. Kadrakhali Khal, an inland arm with another branch called the Damdama Khal forming an inland drainage basin.	1. Bainchapi Khal, a meandering creek.
2. Moni River, a fairly long meandering channels.	2. Guru Khal, a small creek.
3. Pukchara River bifurcated farther north into Chatua and Raidighi Rivers; Raidighi River is again connected to Moni River.	3. Kankalmari – Ajmalmari-Suia Rivers, a large drainage system.
4. The Sibua gang forms loop around Paschim Sripatinagar Island. The southern arm of the loop is connected with Chirapat or Pakhirala creek – a link channel of Kalchera-Curzon creek system. The northern arm of Sibua joins Pukchara through Kuamari Khal.	4. Dhulibasani Gang, an incised tidal creek.
5. Kali Binmari Khal, a small creek between Jagadal River and Thakuran River.	5. Chulkati Gang, a small tidal creek.
6. The Jagadal River, a major tidal channel.	6. Gokultali gang separates the Bulcheri Island in the extreme south.

2.1.10 Mridangabhanga

The Mridangabhanga River originates at Sutarbag. It meets the river Saptamukhi and renamed as Curzon creek. Various creeks namely Chatua River, Nukchara nadi, Kakmara khal, Kuemuri creek are on the east and it joins with the Saralda, Gobadia River and Barchara river on the west. Mridangabhanga River seems to be a large tidal river both by its width and length. The Sutarbag – Mridangabhanga River seems to be a large tidal river. The Sutarbag – Mridangabhanga – Calchara – Curzon creek constitutes one continuous estuarine arm of which the Curzon creek is almost an estuary itself.

2.1.11 Gobadia

Kaloya River is renamed as Gobadia River in the east. The Saralda Khal links the Mridangabhanga with the Gobadia River. Monkedoani khal emerging out of Gobadia River meets the Ghugudanga khal flowing through the meandering course with the name of Gandakata Khal. Barchara river branches off the Gobadia River on the east, which connects the waterways between Patharprotima and Kakdwip P.S.

2.1.12 Saptamukhi

Combination of numerous creeks namely Thikara khal on the north, Banstala khal and Ghugudanga Khal on the South form the river Saptamukhi. The major branches on the west are the Kalnagini khal, Ghugudanga Gang and Blind creek, Hatania – Doania creek, Patibunia creek, Edward's creek, Chandanpiri khal and Bakkhali River around Fraserganj. From north to south the Banstala -Ghugudanga gang – Saptamukhi west may be considered one continuous arm (Fig. 2.1). The Saptamukhi estuary with large width but shallow depth is studded with numerous mid channel bars emerged from the riverbeds.

2.2 Morphodynamics of the Tidal Rivers

Sedimentary processes in the tidal channel flow are associated with the depositional and erosional features like development of large bedforms, sediment waves, crevasse splays and slump scars (Ginsberg and Perillo 2004). The river Channels in the Hugli Matla delta complex of the Sunderbans of NE India experiences a semi-diurnal spring tide range above 5 m near mouth. The tidal prism covers almost the entire length of the rivers. Most of the sediments of the rivers are cycled by tidal currents under extreme paucity of fresh water discharge except for the monsoonal months from the headland. The intertidal sand flats are ornamented by bedforms of various kinds and scales, which help understanding the hydrodynamic condition of the river.

The important sub-environments of deposition along the river basin include (i) marginal linear bars similar to levee deposits built by interaction of ebb-and flood-tidal currents (ii) crescentic point bars in the meandering stretches (iii) mid-channel shoals (flood-tidal delta) formed by accumulation of sediment on the land-ward side of the river by flood tidal currents, and (iv) river mouth bar or outer shoals (ebb-tidal delta) formed by accumulation of sediments seaward by ebb-tidal currents. The distribution of mangrove swamps in relation to space and time has been studied in the context of the above environmental realms.

The bedforms like ripples, mega ripples, sand waves that occur in the ebb- and flood-oriented fields closely resemble that described from the macrotidal Bay of Fundy, Canada and Loughor estuary, UK. Certain specific stretches of present day erosion and deposition simultaneous with neo-tectonic subsidence have been identified along the river course.

2.3 Geomorphic Sandbodies

The geomorphic sand bodies along the rivers of Sunderbans conform to that of a meso-macro-tidal river (Terwindt 1988). The various geomorphic units are as follows:

(i) Marginal levees built by interacting flood and ebb currents: These linear sand bodies are discontinuous along the river length. The upper parts of these levee deposits, during neap cycles, undergo desiccation and are often marked by mud cracks and marshy vegetations.

(ii) Crescentic point bars in meander bends with features indicating flow reversals by ebbing and flooding waters over them: These point bars are totally submerged at high tides and are modified by waves and tidal currents. By contrast, the point bar of a flood plain river is submerged during flood. In this respect, the morphological property of point bars of a tidal channel is distinctive from that of a flood plain river (Friedman and Sanders 1978).

(iii) Mid-channels bars (flood-tidal delta) built by accumulation of sediments on the landward side of the river by flood-tidal currents: Mapping reveals a gradual migration of the bar towards landward side of the river. The migration rate is measured to be approximately 1 km in 60 years (Fig. 2.3).

(iv) River mouth bar or outer bar (ebb-tidal delta) built by accumulation of sediments seaward by ebb-tidal currents: The Bulcheri island stands as a classical zone of sedimentation as a mouth bar of Thakuran River; the southern part of which faces the wave effects of the Bay of Bengal.

2.4 Bedforms: Indicators of Hydrodynamics

Certain important bedforms of the lower flow regime (Simon and Richardson 1962) have been studied from the river channels in the Sunderbans (Table 2.1). These bedforms with their geomorphological attributes act as diagnostic of the flow velocity, water depth and direction of movement of ebb and flood tidal currents. Many of these surface sedimentary structures are unlike those produced in the laboratory. They occur in the ebb and flood oriented fields of the channel and are comparable in many respects to that of the macrotidal Bay of Fundy, Canada (Dalrymple et al. 1978) and Loughor estuary, U.K. (Elliott and Gardinar 1981). Table 2.1 shows the bedforms types and their brief descriptive. The large-and small-scales of the bedforms are defined after Allen (1968).

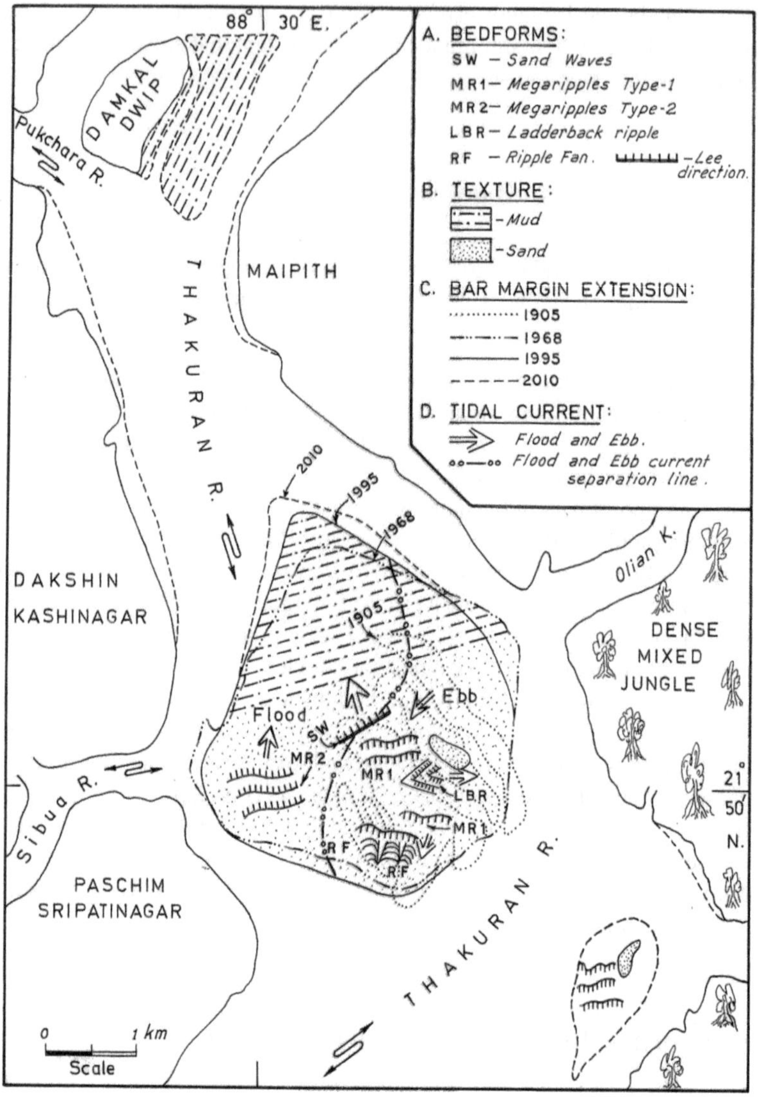

Fig. 2.3 Generation and migration of the mid channel bar in the Thakuran river bed

2.5 Riverine Geomorphic Environments

Many geomorphic environments of deposition and erosion along and across these watercourses of Sunderbans are recognisable. These include the following:

(i) Intertidal to supratidal riverbank (Natural levees)
(ii) Intertidal point bars

Table 2.1 Description of bedforms of the river channels of Sunderbans

Small-scale bedforms	Large-scale bedforms	Modified bedforms of intermediate and large –scale
1. Ripples		
(i) Straight-crested symmetrical	(I) Sand waves-low with extension for tens of meters	(I) Skewed Spurs; sharp crested spurs from slip face of type/ megaripple: produce from helical flow cells in front of slip face
(ii) Straight-crested asymmetrical	(II) Megaripples: Type I (after Dalrymple et al. 1978)-low height, low span, devoid of scour pits.	(II) Ripple fan: continuous arcuate crest line fanning away from slip face of Type I Megaripple
(iii) Lingoid	(III) Undulatory Megaeipple (after Allen 1968)-higher height, variable span, sinuous crest scour pits present.	(III) Dissected channel
		Microdelta-(I) Sandy
		(II) Muddy
(iv) Multiple-crest wave ripples		
(v) Capped-off ripples		
(vi) Ladder-back of complex nature		
(vii) Longitudinal		
(II) Rhomboid marks		
(III) Rill marks		
(i) Conical		
(ii) Bifurcation		
(iii) Branching		
(iv) Sand ribbons		

(iii) Intertidal sub-tidal mid-channel bars
(iv) River mouth bars
(v) Intertidal swash platform

2.5.1 River Banks (Natural Levees)

Both banks of the river systems are generally characterized by fringing mangroves. As a result, the nature of stratifications of the banks or that of the natural levee deposits is often heavily disturbed by roots and pneumatophores of mangroves and by the churning effect of burrowing organisms. In places of massive erosion, the banks have been fortified by construction of embankments. As a result mangroves are truncated in the intertidal zone. Landward, the intertidal area behind the embankments is occupied by marsh vegetation (Bhattacharya 1999). The topmost portions of the bank deposits register evidences of long exposures and contain desiccation cracks and salt-encrusted surfaces in patches. The following features characterise the supratidal levee deposits:

(i) Marshy areas are characterised by marsh lamination with fine obscured laminae of silt and clay profusely disturbed by roots of marsh vegetation.
(ii) Rapid thinning of sand laminae and dominance of mud in the upward direction bank deposits are evident in many bank sections.
(iii) Plant remains and plant roots constitute a substantial component of bank deposits, particularly towards their upper portions.
(iv) Levee surface sometimes shows salt-encrusted patches due to evaporation of saline water pools formed during overbank flooding.
(v) Suspensional mud deposits on the topmost part of the river banks exhibit mud cracks or curly mud cracks and indicate subaerial exposure. Curly mud cracks are often associated with areas having algal matting.

In vertical sections the bank deposits are separable into three units based on colour from the top to the mean low water line. The bank top is dominated by mangrove species *Phoenix paludosa Ceriops decandra and Excoecaria agallocha.* The top stratum mud deposits (upto 2 m thickness) are rigorously disturbed by its penetration of mangrove roots up to a depth of 1 m or so. The immediately underlying unit (generally of 1.5–2 m, thickness) is profusely bioturbated by crab burrows and penetration of mangrove rootlets. The muddy burrow walls and root affected mud cliffs display clear evidences of oxidation marked by brown colour of sediments. The lowermost unit of the banks (up to 2 m thickness) is less affected by bioturbational disturbance. In majority of the cases, the topmost unit has an erosional contact with the underlying intermediate unit, whereas no such erosional contact is recognizable between the intermediate and the lowermost units excepting a distinct colour variation.

2.5.2 Point Bar

Most of the Sunderbans river mouths are macrotidal (tidal amplitude >4 m). The tidal regime falls farther upstream to become mesotidal (tidal amplitude 2–4 m) to microtidal (tidal amplitude <2 m) near their sources. The rivers are generally sinuous to meandering and are characterised by a limited number of point bar deposits where deposition takes place laterally on the convex sides of the meander bends (Fig. 2.4). The sandy point-bar deposits in the macrotidal stretch are gradually replaced landward by mesotidal gently inclined laterally accreting deposits identical to large-scale "epsilon cross stratification" or ECS of Allen (1963) or "longitudinal cross-bedding" of Reineck and Singh (1980). These stratifications are gently inclined towards the river direction. Thickness of these cross-stratified units ranges from 2 to 4 cm. The inclination of the cross-beds averages 15° and

Fig. 2.4 Crescentic point bar at the meander bend of the river

varies between 10° and 20°. ECS differs in origin from the cross-stratifications produced by migration of ripples and megaripples (Reineck and Singh 1980, 1986). Lateral shifting of point-bars facilitates epsilon cross-stratifications to run parallel to the current direction (Smith 1988).

The point-bars are generally falcate to crescentic in shape (Fig. 2.4) and extend for 0.1–6 km in length and 0.2–1.1 km in breadth. Maximum height of point-bars from mean low water line is variable from 2 to 4 m. The length breadth ratio (L/B) of the point-bars nicely conforms that of the meander bends. Sometimes, two to three point-bars coalesce along their lengths to give rise to their extremely elongated shapes.

2.5.3 *Mid-channel Bar*

These geomorphic bodies are quite conspicuous in the central part of most of the riverbeds of Sunderbans. They occur as isolated elliptical depositional shoals extended along the river channel axis. The actual extension of the bars and their limits were studied during ebb tides only. The mid-channel bars generally maintain a gradual upward convexity having lengths ranging from 2.1 to 4.74 km and breadth 1.0 to 3.7 km (e.g. Thakuran River). The height ranges from 2 to 4 m above mean low water level. They migrate upstream with the movement of flood tidal currents along the main flood channel. This upstream migration of a mid channel bar (at Paschim Sripatinagar) in the Thakuran river, has been estimated to be 1.6 m per year for the period of over 25 years (Fig. 2.3).

Many of these mid-channel bars are traversed by meandering creeks and are seen to be active during the low tides. The mid-channel bars that occur farthest from the sea face are generally muddy, whereas, those nearer the sea are mostly sandy. Some mid-channel bars that occur in the middle stretches of the rivers of Sunderbans exhibit excellent grain size differentiation from their northern (landward) to the southern (seaward) extremities. Generally the southern halves of these bars are sandy and the northern halves are muddy in nature. The sandy halves often show a distinct separation of the flood and ebb currents as bedforms representative of these two currents are more clearly developed in the cohesionless sand grains.

2.5.4 *Swash Platform*

These are generally crescentic bars of variable dimensions near the mouths of almost all the major rivers of Sunderbans. These sandy platforms experience wave swash and backwash within the limits of intertidal zone and eventually behave like swash platform (Oertel 1972) where the accretionary sandy bar sediments are reworked by wave action. Thus swash bars exhibit features like that of sandy beaches. Unlike seaward dip common to the beaches, inclination of the swash bar surface is towards the funneled estuaries. The elongation axes of the swash platforms are almost perpendicular to the shoreline.

Morphologically, the swash platforms show gently dipping surfaces (4–5°) with minor undulation on higher topographic areas. These undulations are due to the presence of the backwash ripples, which characterise the upper foreshore zone of most beaches. The maximum advance of wave swash is marked by swash marks made up of wood pieces, weeds and mangrove leaves. The mid intertidal zone of the platforms shows major undulations due to the presence of megaripple trains. Some convex upward ridges on the platforms are ornamented both by large-scale and small-scale rhomboid marks. The lower most stretch of the platforms is characterised by abundant current crescents, rill marks and small-scale wave ripples. At the distend of the platforms where they merge into the muddy mangrove banks, the sandy swash platform are underlain by cohesive mud that supports luxuriant mangrove vegetation, particularly *Phoenix paludosa* forests, the suitable abode for the famous Royal Bengal Tiger of Sunderbans.

2.5.5 *River Mouth Bar*

In all the tidal channels and river courses, unconsolidated sand bodies occur as mouth shoals or river mouth bar (Ginsberg and Perillo 2004). The river mouth bars of Sunderbans are the most prominent sites of sedimentation at the confluence of the major rivers and the Bay of Bengal. Some of the river mouth bars of Sunderbans include Jambu Island, Lothian Island, Bulchery Island, Hallidey Island, Bangaduni Island and Dalhousie Island from the West to the east at the mouths of the major rivers. The upper surfaces of all these islands remain exposed even at high tides throughout the year. Majority of these islands are truly the ebb-tidal deltas (Hayes 1976) and are placed on the ebb-channels. The northern parts of these islands are generally protected from wave action and are under the influence of tidal currents. The southern parts, on the other hand, face the Bay of Bengal and experience strong wave attacks. The river mouth bars originate where the ebb flows are countered by the incoming refracted waves from the open sea (Hubbard et al. 1979).

In many occasions, the intertidal and supratidal zones of these islands are clearly discernible. The intertidal zones exhibit dove-tailing of sand and mud as a result of local variations of energy levels. The supratidal sands are very fine and well sorted. These islands are commonly dissected by a number of tidal creeks of sinuous nature. These creeks are generally ephemeral and are subjected to change their courses with time. Natural creek sections often reveal alternations of sand and mud layers of variable thickness ranging between 10 and 45 cm (eg. Jambu Island, Bulchery Island). Accretion of the Jambu Island has been observed along the north-east side of the island although almost entire southern part of the island has been undergone erosion over a period of 25 years.

It is evident from the study that each geomorphic zone has characteristic sand body geometry to be identified by certain important parameters as dimension of sand bodies, grain size, bedforms, stratifications and facies sequence (Coleman et al. 1988; Terwindt 1988; Banerjee 1989). Despite some overlapping, these parameters present clear distinctiveness of each environment (Table 2.1).

2.6 Tidal Creek Systems

Intertidal mudflat on point bars bordering the river systems are very commonly dissected by tidal creeks. The creeks are generally 10–25 m long and 2–10 m wide. They are sinuous to meandering in pattern and act as pathways for both flood and ebb flow depending on alternation of tidal currents (Fig. 2.5). In many occasions, interlocking ridges and spurs along meandering paths characterise the creek morphology. Various morphological features are recognisable in creek sedimentation.

Mudflats predominate in the low energy areas of suspensional depositions, particularly in the middle to the upper stretches of the rivers belonging to the meso-macrotidal settings. The areas of mud deposition include certain portions of mid-channel bars, point-bars and margins of tidal creeks that often intersect the intertidal flats (Fig. 2.5). The mudflats occur covering few sq m to hundreds of sq m areas with thickness of mud blankets ranging from few centimeters to less than a meter. Fine rhythmic laminations of silt and clay characterise the deposits. Lenses of silty sand of 2.12 mm thickness occasionally intercalate the mud deposits. Physical and rheological properties (viscosity, plasticity and thixotropy) are known to be governing factors for erosion and resuspension of fine-grained sediments in tidal river and estuarine environments (Faas 1981; Amos 1995). Generally the tidal mud deposits are soft, grey and water saturated. The water content varies at different levels of the accumulated mud.

Fig. 2.5 A typical meandering tidal creek of Sunderbans

Granulometric properties reveal a range of silt content varying from 49.40 to 99.57 %, clay content from 0.14 to 34.41 % and mean size from 4.45 to 7.7 phi. Organic matter content is always higher in the upper mudflats. Mineralogically, the silty fraction is composed of 85.90 % quartz. 5.6 % feldspar, and rest mica. Generally, micas concentrate more towards the upper and landward stretch of the river with low tidal energy conditions. The dominance of quartz (with less feldspar) characterise a siliciclastic nature of the mudflats. Micas, because of their platy or flaky habit are very sensitive to only slight changes in the energy conditions. Biogenic components of the mud deposits rarely exceed 10–15 % and are composed of fine mangrove debris, fragmentary and entire shells of crabs, gastropods, bivalves and foraminifers. Fecal materials of these animals also constitute a part of sediments.

2.6.1 *Mud Ridges*

Small mud ridges normal to creek axes are developed on steeply inclined (40–45°) creek slopes. Freshly deposited mud on creek slopes suffers differential erosion because of drainage of water through intermittent rills. The uneroded portions remain as mud ridges on creek slopes. The ridges are 12–15 cm broad, 5–10 cm high and 40–45 cm long. They usually occur in complex groups and show less bifurcation compared to the types described by Reineck and Singh (1980). Many of these mud ridges are also produced by liquefaction of mud already settled along creek slopes.

2.6.2 *Mud Microdelta*

Microdeltas are formed by the accumulation of mud on the creek floor. These are triangular in plan and wedge shaped in longitudinal section. Their surface is slightly convex upward. The microdelta fronts are lobed and prograding. They occur as solitary bodies and do not occur in trains. They are 20–25 cm wide, 15–20 cm long and 5–10 cm thick. Mud transported along the gently sloping creek margins is responsible for micro delta formation. Gullies or rills initiate the pathway of mud transportation.

2.6.3 *Mud Pellets and Mud Lumps*

Mudflats around creek margins are generally strewn with mud pellets and mud lumps (Dalrymple et al. 1991). The pellets are flat subelliptical bodies with axial lengths ranging from 2 to 5 cm. Shape analysis of the pellets reveals that the pellets mostly plot on the prolate and oblate sectors in the Zingg's shape class. The exposed portions of mud banks undergo erosion due to undercutting by tidal currents. As a result, chunks of mud of variable dimensions collapse from the mud banks. These mud chunks later break down into smaller pieces.

2.6.4 *Wood Clumps on Mudflats*

Clumps of mangrove wood are often associated with the muddy sediments. These wood trunks or stems, 1–3 mm long and 2–30 mm diameter drift with tidal currents and get stuck into the mud at lower flow velocity. Their presence in the mudflat is a definite indication of their tropical environment origin. These wood trunks are derived from the eroded bank material sustaining mangrove forests.

2.7 Embankments

The natural levees and the banks of the rivers of Sunderbans are often affected by severe erosion and thus need protection by some artificial means. Embankments of about 4,250 km length protect the margin of Sunderbans Rivers (Fig. 2.6).

Fig. 2.6 Earthen embankment typical of Sunderbans noted with basal destruction

The embankments are mostly brickworks over earthen dams made to save the hinterlands from saltwater menace. Although important as a socio-economic measure these constructions have short- and long term ecological impacts over the area (Poff et al. 2007). Large scale construction of concrete embankment all over the 54 islands out of total 102 islands of the Sunderbans which is the dwelling denizen to about five million people has the major impact over the fragile ecology of this mangrove lands. The concrete embankment causes the barrier to the unique mangrove dwelling fauna at the natural levee and river flood plains resulting detrimental effects on the mangrove ecosystems. The absence of a particular fauna leads to a definite gap in the food chain and ultimately the structural frame work of the food web becomes disturbed. The river banks of the Sunderbans is not only the habitats of hundreds of non-chordates like pea crabs, fiddler crabs, sea pens or sea anemones, but is the denizen and breeding grounds of the birds like kingfishers and sea gulls. The concrete embankments are not feasible in protecting the embankments that are seen eroded away by the tidal actions in the Mahishani Island of the Namkhana block. Polypropylene sheets used for the concretization of the embankment will be harmful to the estuarine environment as the polypropylene is not a bio-degradable ingredient. As a result the construction of the concrete embankments using polypropylene intervene the habitat structure of the mangrove fauna resulting detrimental impact over this dynamic ecosystem.

The following physiographic and morphological changes of the river basin have been noticed because of these large-scale constructions.

(i) Embankments prevent tidal flushing in the intertidal areas and so a large volume of sediments carried by the river is deposited on the riverbed instead of getting dispersed in the inter-channel areas. As a result, the riverbed gets choked up rapidly causing the natural water flow to change its direction. This can be considered as the main reason for the ever-changing course of the Sunderbans Rivers with a consequent rise of the riverbeds.

(ii) Embankments impose restrictions to the spill areas of the river. Eventually some water courses have dried up and degenerated beyond replenishment (Sharma 1994).

(iii) Embankments restrict flourishment of mangroves in the inter-channel areas (Woodroffe 1999; Bhattacharya 1999). Further, growth of fringing mangroves along the banks of the Sunderbans Rivers is hindered to a great extent because of such constructions.

2.8 Summary

The rivers including Ganga-Brahmaputra-Meghna (GBM) are the principle contributors to the depositional environment of the Bengal basin. The river load of Ganga-Brahmaputra is estimated to be 1 billion tons at river gauging stations-300 km inland of the coast of the Bay of Bengal (Milliman and Syvitski 1992) as calculated on the basis of geochronological and geophysical investigations. A sediment budget, of that carried load by these two rivers, indicates the following partitioning: (i) 30–39 % is sequestered landward of the ocean in the huge flood plain (Goodbred and Kuehl 2000; Allison 1998); (ii) 21 % is deposited in the top set beds of the subaqueous delta near the river mouth (Allison 1998), and seaward progradation of the subaqueous delta front incorporates 20 % of the river load (Michels et al. 1998) and (iii) the remaining 20–29 % probably reaches the Swatch of No Ground (Goodbred and Kuehl 2000).

Allison (1998) documents net erosion of the Sunderbans shoreline, increasing to the west, where 3–4 km of retreat has occurred since 1840. The alongshore difference in erosion rates suggest the western Sunderbans i.e. of Indian part is sediment starved, either by the eastward migration of the river mouths or by decreased Ganga sediment delivery via local distributaries. Whatever the case, shore line erosion and saline intrusion may also be compounded by regional subsidence (tectonic and compaction) process. The coastal geomorphic features of the recent times are seen with various sedimentary environment of estuarine river deposition in the lower part of the Ganges delta surrounding the total area of Sunderbans. The different geomorphic units include the river flood plains, marshes and swamps, estuaries, beaches, tidal flats and the river mouth bars. The sedimentary facies have also been clearly delineated. The granulometric character shows a dovetailing of silt sand and clay facies in the river flood plains. The marshes and swamps are characterised by sandy and clayey silt facies and the dune with fine clear sand and silt facies, whereas, the intertidal beaches and off shore bars, the typical sandy (medium to fine) facies. The hinterland in the back dune areas represents silty clay facies. There is, however, a gradual fining of grains as one proceeds towards the off shore region. Tidal creek and mudflat sedimentations are of two main types in the Sunderbans (i) lateral sedimentation and (ii) vertical sedimentation. Lateral sedimentation takes place on meandering tidal creeks and gullies, and results in dipping laminae. Deposition here is rather rapid. Vertical sedimentation, on the other hand, is found on the higher parts of the mudflats, above the influence of tidal creeks. Here sedimentation is rather slow and yields horizontal laminae. Both inclined and horizontal mud laminae interfinger near edges of the raised mudflats.

References

Allen JRL (1963) The classification of cross-stratified units with notes on their origin. Sedimentology 2:93–114

Allen JRL (1968) Current ripples: their relation to patterns of water and sediment motion. North Holland Publishing Company, Amsterdam

Allison MA (1998) Geological framework and environmental status of the Ganges- Brahmaputra delta. J Coast Res 14(3):826–836

Amos CL (1995) Siliciclastic tidal flats. In: Perillo GME (ed) Geomorphology and sedimentology of estuaries, Developments in sedimentology, 53., Elsevier, Amsterdam, pp 273–306

Banerjee I (1989) Tidal structures in the Glauconitic Sandstone, Countess Field, Southern Alberta, Canada. In Modern and ancient examples of clastic tidal deposits – a core and peel workshop. In: Reinson GE (ed) Canadian Society of Petroleum Geologists, second international research symposium on clastic tidal deposits, Calgary, Alberta, pp 89–97

Bhattacharya A (1999) Embankments and their ecological impacts: a case study from the tropical low-lying coastal plains of the deltaic Sunderbans, India. In: Vollmer M, Grann H (eds) Large-scale constructions in coastal environments, Springer, Berlin, pp 171–180

Coleman SM, Berquist CR Jr, Hobbs CH III (1988) Structure, age and origin of the bay mouth shoal deposits, Chesapeake Bay, Virginia. Mar Geol 83:95–113

Dalrymple RW, Knight RJ, Lambiase JJ (1978) Bedforms and their hydraulic stability relationships in a tidal environment, Bay of Fundy, Canada. Nature 275:100–104

Dalrymple RW, Makino Y, Zaitlin BA (1991) Temporal and spatial patterns of rhythmite deposition on mudflats in the macrotidal, Cobequid Bay–Salmon River estuary, Bay of Fundy. In: Smith DG, Reinson GE, Zaitlin BA, Rahmani RA (eds) Clastic tidal sedimentology. Memoir Canadian Society of Petroleum Geology, 16, pp137–160

Elliot I, Gardiner AR (1981) Ripple, megaripples and sandwave bedforms in the macrotidal Loughor Estuary, South Wales, UK. Spec Publ Int Assoc Sedimentol 5:51–64

Faas RW (1981) Rheological characteristics of Rappahanock estuary muds, Southeastern Virginia, USA. Spec Publ Int Assoc Sedimentol 5:505–515

Friedman GM, Sanders JE (1978) Principles of sedimentology. Wiley, New York

Ginsberg SS, Perillo GME (2004) Characteristics of tidal channels in a mesotidal estuary of Argentina. J Coast Res 20(2):489–497

Goodbred SL Jr, Kuehl SA (2000) The significance of large sediment supply, active tectonism, and eustasy on margin sequence development: late Quaternary stratigraphy and evolution of the Ganges – Brahmaputra delta. Sediment Geol 133:227–248

Hayes MO (1976) Morphology of sand accumulation in estuaries: an introduction to the symposium. In: Cromn LE (ed) Estuarine research, vol II. Geology and engineering. Academic, New York, pp 3–22

Hubbard DK, Oertal G, Nummedal D (1979) The role of waves and tidal currents in the development of tidal inlets sedimentary structures and sand body geometry: examples from North Carolina, South Carolina and Georgia. J Sediment Petrol 33:604–608

Michels KH, Kudrass HR, Hubscher C, Suckow A, Wiedicke M (1998) The submarine delta of the Ganges – Brahmaputra: cyclone dominated sedimentation patterns. Mar Geol 149:133–154

Milliman JD, Syvitski JPM (1992) Geomorphic tectonic control of sediment discharge to the ocean: the importance of small mountain rivers. J Geol 100:525–544

Oertel GE (1972) Sediment transport on estuary entrance shoals and the formation of swash platform. J Sediment Petrol 42(4):858–863

Poff NL, Olden JD, Merritt DM, Pepin DM (2007) Homogenization of regional river dynamics by dams and global biodiversity implications. Proc Natl Acad Sci U S A 104:5732–5737

Reineck HE, Singh IB (1980) Depositional sedimentary environments. Springer, New York

Reineck HE, Singh IB (1986) Depositional sedimentary environments – with reference to terrigenous clastics. Springer, New York

Sharma BC (1994) Lower Sunderbans. The Asiatic Society Publ., Calcutta

Sharma AP (2010) Coastal zone vegetation in India with reference to mangroves and need for their conservation. In: Sunderbans: issues and threats. CIFRI (ICAR), Kolkata, pp 9–40

Simon DB, Richardson EV (1962) Resistance to flow in alluvial channels. Trans Am Soc Civ Eng 127:927–953

Smith DG (1988) Modern point bar deposits analogous to the Athabasca Oil sands, Alberta, Canada. In: de Boer PL, VanGelder A, Nio SD (eds) Tide – influenced sedimentary environments and facies. D. Reidal Publishing Co., Holland, pp 417–432

Terwindt JHJ (1988) Paleo-tidal construction of inshore tidal depositional environments. In: de Boer PL, van Gelder A, Nio SD (eds) Tide-influenced sedimentary environments and facies. D. Reidel Publishing Company, Dordrecht, pp 233–264

Woodroffe CD (1999) Response of mangrove shorelines to sea level change. Tropics 8(3):159–177

[illegible]

Chapter 3
Brackish Water

Abstract The rivers of Sunderbans along with its numerous creeks, tidal inlets and estuaries forming an anastomising network together with submerged swamps and marshes, constitute its low saline brackish water body. The brackish water of Sunderbans have the potential for the growth of mangroves and is the breeding ground for large varieties of fin fish and shell fishes. Neo tectonic movement has resulted in the change of courses for the river Hooghly which impeded fresh water flow towards Sunderbans. Changes brought by regional and global processes, climatic changes etc. have resulted in increasing trends of water and soil salinity, loss of biodiversity, degrading soil and water quality which are not congenial for this vulnerable ecosystem.

Keywords Sunderbans • Tidal rivers • Creeks • Salinity • pH • Dissolved oxygen • Alkalinity

Sunderbans (of both India and Bangladesh) is the single largest chunk of deltaic mangrove zones of the planet. Covering an area of 9,630 km^2, Indian Sunderbans is a part of the world's largest fluvio-marine Ganga-Brahmaputra delta at the confluence of the Bay of Bengal (Das and Dutta 2004; Das et al. 2004). It is a well-known Biosphere Reserve of India intersected by numerous tidal creeks and rivers and is famous for its mangrove ecosystem in the tropical region.

A large number of tidal rivers, creeks, tidal inlets, estuaries and network of intricate minor salt water courses along with submerged swamps and marsh together constitute the water body of the Sunderbans. The sea-land interphase mangrove ecosystem comprises of about 55 % forest land and 45 % water spread area in the forms of tidal rivers, creeks, inlets and vast estuarine mouths of the river Hooghly, Muriganga, Saptamukhi, Thakuran, Matla, Gosaba and Haribhanga. The rivers (31 in numbers) of Sunderbans along with its numerous tidal creeks and canals render this entire area like a criss-cross network. Water courses including rivers, creeks, tidal inlets, salt water courses are characterised by low saline brackish water where saline waters get diluted with the freshwater only from precipitation. Waters of these estuarine rivers and creeks have much potential for the growth of the mangroves as well as for nurturing vast quantities and varieties of both fin fish and shell fishes.

G.K. Das, *Estuarine Morphodynamics of the Sunderbans*,
Coastal Research Library 11, DOI 10.1007/978-3-319-11343-2_3

Sunderbans provides low lying deltaic plains with meso-macrotidal amplitude that helps generating the luxuriant growth to mangroves. A large number of mangrove species and their associates or back mangroves are characteristically salt-tolerant. A few species are grown at high salinity substrate whereas maximum mangrove species prefer the low salinity region. Some mangrove species are inundated twice daily in the semidiurnal tidal regime during flood tide. Soil and water salinity including other physico-chemical parameters are important factors in controlling the growth and occurrences of mangroves in Sunderbans areas.

3.1 Physico-chemical Properties

Most of the mangrove species prefer low saline substratum. The mangrove species inundated twice daily by tides prefer soils with hyper-salinity whereas species which occur in the central portion of islands prefer their habitat with low salinity. Mangrove species diversity and richness depend upon the preferred salinity of respective mangrove species. Mangrove zonation occurs in response to the salinity of both soil and water. Water being the primary requisite to support aquatic life its physico-chemical properties is responsible for maintaining the aquatic environment and the water quality parameters are likely to influence the growth of the organisms and thereby, productivity of the ecosystems.

Changes in the local hydrology are the most important contributory factors attributable to the changing scenario of mangrove ecosystems. Changes in river courses impede fresh water flow towards Sunderbans. The scarcity of fresh water supply into the river waters results increasing trends of water and soil salinity, loss of biodiversity and degrading soil and water quality that create threats to this fragile ecosystems. All these changes as a result of regional and global processes and climatic changes cause the changes in the salinity, pH, dissolved oxygen, temperature and all other changes in the physico-chemical parameters in the brackish waters of the Sunderbans. Mangrove distribution is limited worldwide depending principally on temperature and salinity variations of the estuarine river waters.

3.1.1 pH

In the river waters of the Sunderbans pH is an important factor in the mangrove ecosystem to influence the physico-chemical reactions as well as many biological activities. pH activates microbial growth rate which in turn influences the bio-degradation of mangrove litters and other microbial mediated biochemical reactions. Most favourable pH condition is 6.0–8.0 for microbial degradation of biodegradable materials. pH measured in the tidal courses and inlets ranging between 7.3 and 7.7 is within the permissible limit in the waters of Sunderbans (Table 3.1) for the growth of mangrove vegetations.

Table 3.1 Water sample analysis in the rivers of Sunderbans

Flood (F) Ebb (E)	Sample location	pH	Temp (°C)	DO (mg/L)	Salinity (ppt)
E.	Hogal R.	7.5	21.2	7.2	18
E.	Bidya R.	7.7	21.5	4.9	20
F.	Sajnekhali	7.5	21.4	3.8	23
F.	Gumdi	7.3	18.8	4.4	25
F.	Bharani-2	7.3	19.4	4.5	25
F.	Pirkhali	7.3	19.7	4.0	25
F.	Gazikhali	7.4	20.0	4.4	25
E.	Panchamukhani-1	7.5	20.1	4.6	25
E.	Dobanki	7.4	20.4	4.7	25
E.	Panchamukhani-2	7.4	20.6	3.5	25
E.	Jharkhali	7.5	20.8	3.6	25
E.	Netidhopani	7.5	21.1	3.3	25
E.	Chamta	7.4	21.4	6.8	21
E.	Deul Bharani	7.3	21.1	6.0	21
E.	Chora Gazi	7.3	20.7	5.2	22
E.	Bara Gazi	7.4	20.3	5.2	23
E.	Andharbani	7.4	20.2	4.0	23
E.	Sundarkhali	7.4	19.1	4.6	24
E.	Banbibi Bharani	7.3	18.1	3.3	24
E.	Sudhanyakhali R.	7.4	18.7	3.4	25
F.	Pakhirala	7.6	18.3	3.0	24
F.	Pakhir Khal	7.3	18.2	3.0	23
F.	Guritana Khal	7.4	19.9	5.2	23
F.	Sakuntala Khal	7.3	20.4	3.3	23
F.	Sudhanyakhali	7.4	20.5	4.0	23

3.1.2 *Temperature*

Water temperature of the river depends mainly on climate, sunlight and depth of water. All metabolic and physiological activities and life processes such as growth, feeding, reproduction and movement of the aquatic organisms are influenced by water temperature. The solubility of oxygen in water is inversely related whereas other solutes are directly related to water temperature. The rate of decomposition of organic matter of bottom soil is also regulated by temperature to a great extent. In tropical tidal estuarine environment of the Sunderbans water temperature undergo a wide diurnal and seasonal variation, which normally do not show adverse effect.

Water temperature is one of the most important physical factors, which influence the growth of aquatic organisms in the estuarine systems. The mean water temperature decreases with the channel length from the source point towards the river mouth in the deltaic Sunderbans. As the amount of suspended solids decreases and depth of the channel increases, water is observed to retain lesser amount of heat. Water temperature varying between 18.1 and 21.5 °C is well within the range. Mangroves cannot tolerate cold temperature less than 20 °C for a continuous period. For these reasons, mangroves are found only in the tropical and sub-tropical region including Sunderbans. The shallow depth estuarine brackish water temperature is always a few degrees higher than the adjacent coastal waters of massive influence.

3.1.2.1 Temperature Variations

Temperature of the flood-tidal river waters of Sunderbans is recorded higher than that of the ambient temperature in the month of December during winter time. Increased water temperature is a local phenomenon in the Sunderbans region and may be the resultant heat of mixing of the silt particles heavily and also the different chemicals occurred in the vortex of sediment-laden river waters particularly due to asymmetric bottom topography during flood tide situation when waters vigorously enter into the upstream of the river course coming out of the Bay of Bengal (Das 2012).

It happens that the sediment laden river waters of Sunderbans carry a large content of silt particles, which results in high TDS values (Table 3.2). Innumerable vortices in the river water appear due to irregular bottom topography and helicoidal

Table 3.2 Analyses of the water samples collected during winter time from the rivers of Sunderbans

Sl. no	Sample location	Atoms. temp (°C)	Water temp (°C)	pH	Salinity (ppt)	TDS (ppm)	DO (mg/L)	Flood (F)/ Ebb (E)
1	Sonakhali	27.8	22.4	7.5	12	990	1.8	E
2	Gosaba	24.9	22.6	7.9	14	775	3.9	E
3	Sajnekhali	25.3	22.3	7.9	17	860	3.5	E
4	Gumdi	22.4	24.1	7.8	16	560	3.6	E
5	Dayapur	22.0	18.1	7.9	15	865	2.9	E
6	Satjelia	19.9	20.4	7.9	15	790	2.3	F
7	Chotomollakhali	20.7	23.1	7.9	15	825	2.3	F
8	Malmelia khal	21.4	22.3	8.0	14	750	2.3	F
9	Korankhali	21.4	23.4	7.9	14	810	2.3	F
10	Marichjhapi	22.4	23.1	7.9	14	790	2.5	F
11	Kumirmari	23.3	23.6	7.9	11	695	2.6	F
12	Jhilla R.	22.8	21.6	7.9	6	440	2.5	F
13	Jhilla	24.0	22.8	7.9	7	485	2.8	E
14	Burir Dabri K.	24.2	23.5	7.8	8	655	5.6	E
15	Katuajhuri	24.4	24.1	7.8	7	565	5.7	E
16	Burir Dabri	25.9	21.4	7.9	10	590	2.9	E
17	Pakhirala	21.4	22.2	7.7	16	685	2.4	F
18	Sudhanyakhali	24.8	21.7	7.9	16	800	6.0	F
19	Sudhanyakhali K.	22.1	21.6	7.9	18	775	2.5	F

tidal flows during flood tide in almost all the rivers having meandering bends along their courses in the Sunderbans region. Silt particles undergone mixing with water and the resultant heat of mixing as a result of the kinetic energy from those vortices may make the water temperature warmer than the ambient temperature. During flood tide water molecules move faster resulting in higher temperature. Temperature change is the response of the water substance due to the removal of heat energy for high TDS value of river waters (Das and Dutta 2007). Comparatively warm temperature of the river waters in the Sunderbans during winter may have significant control over the zonation and speciation of mangroves.

3.1.3 Dissolved Oxygen (DO)

The equilibrium solubility of O_2 in water at 25 °C is 2.7×10^{-4} mol L^{-1} (8.7 mg/L). DO is essential for not only the survival of aquatic organisms and growth of primary productions in the waters of Sunderbans but also for various oxidation processes in physico-chemical interactions.

Very high DO of surface water (more than 9 mg/L) is detrimental as high DO will enhance the growth of phytoplankton, zooplankton and bacteria leading to eutrophication which will consume oxygen at a higher rate within water converting the water body into a dead pool of water (Dodds 2009). The eutrophications, which may be detrimental to aquatic life, are compounded by day light excursion in dissolved oxygen due to photosynthesis and respiration. The dissolved oxygen content varied from 1.8 to 7.2 mg/L, the lower values being found in the estuaries of the river Gumdi near Pakhirala and the higher values in the water of Hogol River at Sonakhali (Tables 3.1 and 3.2).

3.1.4 Salinity

Salinity represents the quantity of dissolved salts in a given unit of water and is usually expressed in parts per thousands (‰). Growth and reproduction of the many aquatic flora and fauna and natural food production depend to a large extent on salinity of water. Most of the brackish water species are euryhaline in nature which is related to their osmoregulatory adoptions. Salinity is one of the important parameters as the mangroves are grown in a special type of environment where these plants are flooded with saline waters twice daily. Salinity of coastal waters ranges from 23 to 31 ‰, which becomes less in estuary. Water salinity in the rivers of Sunderbans changes with the seasonal variation (Fig. 3.1). Salinity varies from 8 to 19 ppt from monsoon to pre- and post-monsoon times. It ranges between 10 and 25 ‰ in the surface waters during post-monsoon periods (Tables 3.2 and 3.3) in the Sunderbans. On the contrary, the salinity of open seawater off the mouth of the Rivers varies between 19 and 31 ‰.

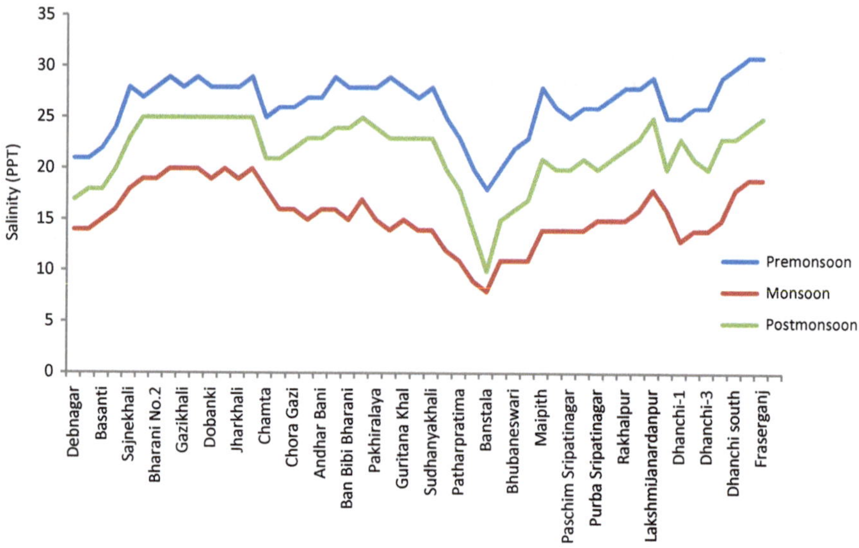

Fig. 3.1 Seasonal variations of salinity in the estuarine river waters of Sunderbans

Table 3.3 Seasonal variations of salinity in the river waters of Sunderbans

Sl. no	Sample location	Salinity in ppt (‰)		
		Pre-monsoon	Monsoon	Post-monsoon
1	Debnagar	21	14	17
2	Namkhana	21	14	18
3	Basanti	22	15	18
4	Gosaba	24	16	20
5	Sajnekhali	28	18	23
6	Gumdi	27	19	25
7	Bharani No.2	28	19	25
8	Pirkhali	29	20	25
9	Gazikhali	28	20	25
10	Panchamukhani-1	29	20	25
11	Dobanki	28	19	25
12	Panchamukhani-2	28	20	25
13	Jharkhali	28	19	25
14	Netidhopani	29	20	25
15	Chamta	25	18	21
16	Deul Bharani	26	16	21
17	Chora Gazi	26	16	22
18	Bara Gazi	27	15	23
19	Andhar Bani	27	16	23
20	Sunder khali	29	16	24

(continued)

Table 3.3 (continued)

Sl. no	Sample location	Salinity in ppt (‰)		
		Pre-monsoon	Monsoon	Post-monsoon
21	Ban Bibi Bharani	28	15	24
22	Sudhanyakhali	28	17	25
23	Pakhiralaya	28	15	24
24	Pakhir Khal	29	14	23
25	Guritana Khal	28	15	23
26	Sakuntala Khal	27	14	23
27	Sudhanyakhali	28	14	23
28	Ramganga	25	12	20
29	Patharpratima	23	11	18
30	Mandir Ghat	20	9	14
31	Banstala	18	8	10
32	Jata	20	11	15
33	Bhubaneswari	22	11	16
34	Saheber Ghat	23	11	17
35	Maipith	28	14	21
36	Harinala Khal	26	14	20
37	Paschim Sripatinagar	25	14	20
38	Chilkamari Khal	26	14	21
39	Purba Sripatinagar	26	15	20
40	Upendranagar	27	15	21
41	Rakhalpur	28	15	22
42	Sridharnagar	28	16	23
43	LakshmiJanardanpur	29	18	25
44	Dakshin Kashinagar	25	16	20
45	Dhanchi-1	25	13	23
46	Dhanchi-2	26	14	21
47	Dhanchi-3	26	14	20
48	Dhanchi-4	29	15	23
49	Dhanchi south	30	18	23
50	Bakkhali	31	19	24
51	Fraserganj	31	19	24

3.1.4.1 Role of Salinity

Sunderbans is the habitat of densely populated mangrove vegetation with a large number of varied species. Drastic changes are revealed with the mangrove zonation on the river banks or natural levees influenced by the changing soil and water salinity from downstream to upstream tidal water courses. Increased salinity causes increased osmotic potential of the water in interstitial soil, which makes the root system difficult for uptaking water. In such condition mangroves adopt different mechanisms through some physiological modifications either by storage, excretion or exclusion of excess salts. Habitats of different mangrove species are generally befitted to their habitats on the magnitude of salinity tolerance. The growth of mangroves may be accelerated with the influence of soil and water salinity (Bandyopadhyay et al. 2001).

High concentration of salinity of soil and water restricts the growth of densely vegetated mangroves. Upstream fresh water discharges which mix up with the saline water results brackish water situation i.e. low salinity condition where mangrove proliferates. It is observed that *Avicennia, Aegialitis, Bruguiera* sp etc. occurred abundantly at the high salinity situations in contrast to *Heritiera, Excoecaria, Xylocarpus, Rhizophora, Ceriops* sp etc. which have been occurred at comparatively low salinity conditions. Even minute variations of salinity induce changing habitats for mangroves of different salt tolerant categories.

3.1.4.2 Impact of Salinity

Growth and productivity of mangroves depend on the climate, hydrology, topography and tidal inundation. But the decreasing rate of respiration may cause damage to the mangrove vegetation if it is inundated for long period even in the fresh water. Rhythmic inundation two times daily in a semidiurnal situation is normal to the mangroves of tropical region like Sunderbans.

High salinity of the river waters round the year may be detrimental to the mangrove community. Long periods of high salinity in an area may trigger complete destruction of mangrove vegetation. Hyper-salinity retards growth of the trees in *A. marina* community (Selvam et al. 1991), lessens biomass of *Bruguiera gymnorrhiza* (Naido 1990). Hyper-salinity in the interstitial water of the saline soil decreases leaf area of mangroves, reduces total NPK and accelerates leaf sap osmotic pressure (Medina et al. 1995). Sometimes mangrove plays a vital role in changing the salinity level in the interstitial water where salt flat is caused due to low rainfall and high evaporation (Ridd and Sam 1996).

It is expected that a rise in sea level would lead to coastal erosion, tidal shift, sea water ingress and degradation of coastal ecosystem. Tidal amplitude will increase and bring changes in the current pattern due to an average sea level rise of 3 mm per year (Brunn and Nayak 1980).The changed current pattern will alter sediment transport, erosion and deposition. The changed tidal regime and sea ingress will affect the present coastal ecosystem and probably, destroy it. In this context the role of salinity over the occurrences, growth and development of mangroves in the Sunderbans region is very important. The mangroves are grown in the substratum of different salinity inundated by brackish waters twice daily.

3.1.4.3 Trend of Salinity Decline

The salinity of coastal waters of the Bay of Bengal has been decreasing slowly as recorded during the period 1994–2011 (Table 3.4).The seawater is diluting very

Table 3.4 Salinity variations in the coastal waters at Bakkhali in the Sunderbans

Month		S	A	L	I	N	I	T	Y	(‰)		
Year	Jan	Feb	Mar	Apr	May	Jun	Jul	Aug	Sept	Oct	Nov	Dec
1994	31	26	26	28	29	31	22	20	20	16	16	31
2011	30	25	26	28	29	31	22	20	19	15	15	30

slowly because the ice sheet of Antarctica has been melting and is releasing an estimated 36 cubic miles of water into the sea each year. The Antarctica is playing a major role in the rise of global sea level at the rate of 3 mm a year (IPCC Report 2007). The melting ice sheet adds water to the oceans that leads to sea level rise as well as decreasing salinity due to dilution of seawater. The coastal water of the Bay of Bengal shows a sharp decline particularly on the month of November in every year (Das 2008). Although the agricultural land run off during October–December has the seasonal contribution for dilution of seawater locally.

3.1.5 *Alkalinity*

Alkalinity of water is its acid neutralizing capacity. Carbonate and bicarbonate are generally the main constituents of alkalinity in the estuarine environments of Sunderbans. Alkalinity is an index of potential carbon dioxide. When the conversion of carbon dioxide to organic matter by photosynthesis is greater than the available carbon dioxide from biotic respiration and organic matter decomposition, required carbon dioxide is drawn from carbonate reserve. Water of Sunderbans Rivers having 50 mg/L or more alkalinity is considered productive. At higher range it does not act as limiting factor. In estuarine waters of Sunderbans total alkalinity vary from 50 to 160 ppm and are directly related with salinity (Anonymous 1987).

3.1.6 *Turbidity*

The turbidity of the estuarine waters is due to suspended solids, which is usually associated with plankton, clay particles etc. suspended in the water. Penetration of light into water phase is determined by turbidity. Transparency plays the major role in primary production i.e. synthesis of carbohydrate, which is a photochemical process energized by light. High turbidity caused by suspended clay and silt particles for long duration make the water body unproductive and sometimes directly affects the aquatic organisms having gills as respiratory organs, causing mortality due to clogging of gills (Anonymous 1987). Secchi disc visibility reading in the river waters of the Sunderbans varies from 37 to 73 cm. More penetration of sunlight

Fig. 3.2 Release of sewage carried by the canal (*left*) into the Bidyadhari River (*right*) at Kulti

causes higher reading of Secchi disc during post-monsoon season whereas, sediment laden waters causes less penetration of sunlight during monsoon period. Releasing domestic sewage carried by the canals from the Kolkata metropolis into the Bidyadhari River increases turbidity of the river water (Fig. 3.2).

3.1.7 Carbon Dioxide

Carbon dioxide plays an important role in the aquatic environment of the Sunderbans by producing bicarbonate from calcium carbonate and by maintaining pH of the water nearly constant through the buffer system of CO_2, $CaHCO_3$, and CaCO3. Most aquatic organisms can survive a wide range of carbon dioxide concentration in the presence of sufficient level of dissolved oxygen in water. At lower level of dissolved oxygen presence of appreciable amount of carbon dioxide affect the uptake of oxygen. Free carbon dioxide above 15 mg/L is considered detrimental to the aquatic organisms.

3.2 Dissolved Nutrients

Aquatic organisms derive their nitrogen requirement in soluble inorganic NH_4 and NO_3 forms and phosphorus in soluble inorganic forms. Waters having inorganic nitrogen and P_2O_5 levels above 0.2 and 0.1 ppm respectively are considered productive.

Table 3.5 Physico–chemical parameters of coastal waters at Bakkhali of Sunderbans

	Dec–Feb			March–May			June–August			Sept–Nov		
Parameters	Stn1	Stn2	Stn3	Stn1	Stn2	Stn3	Stn1	Stn2	Stn3	Stn1	Stn2	Stn3
pH	8.46	8.49	8.45	8.65	8.64	8.52	8.26	8.24	8.26	8.5	8.48	8.5
Temp (°C)	27.5	27.5	27.7	31.5	31.8	31.4	29.8	30	30.1	28.9	28.6	28.7
DO (mg/L)	7.5	7.7	7.8	8.5	8.7	8.9	6.8	7.2	7.6	8.1	7.9	8.5
BOD (mg/L)	4.2	5.3	5.8	3.9	4.4	4.1	5.3	5	3.3	4.1	4.8	3.9
COD (mg/L)	540	556	676	520	560	490	560	496	500	420	468	464
Salinity (‰)	28.5	28.9	28.6	31.6	31.5	31.7	16.5	16.6	16.2	22.5	22.9	22.6
TDS (mg/L)	934	897	676	740	787	780	994	910	887	700	680	720
Turbidity (NTU)	601	757	602	550	600	480	929	808	628	549	550	580
Conductivity (Ms /C)	1.86	1.79	1.97	1.81	1.72	1.87	1.97	1.81	1.77	1.78	1.75	1.81
Bacteria/ml	150	120	86	80	92	76	150	100	96	96	106	80

3.3 Water Quality of Coastal Areas of Sunderbans

Coastal water of the Bay of Bengal at Bakkhali in the extreme southern portion of South 24 Parganas district of West Bengal gets polluted as the analyzed Chemical Oxygen Demand (COD) value and Total Dissolved Solids (TDS) of the coastal water samples cross the permissible limit (Table 3.5). Spatial variations of some physico-chemical properties of the coastal waters of Bay of Bengal at Bakkhali were measured. The mixing of river waters from Hooghly flowing through Haldia industrial belt (basically Chemical detergents, batteries, petrochemicals and plastics) in coastal areas are the primary causes of nutrient enrichment, hypoxia, harmful algal blooms, toxic contamination, sedimentation and other problems that plague coastal water at Bakkhali.

3.4 Summary

The environmental scenario of the present Sunderbans is quite different from that of the last century. The physico-chemical characteristics of the brackish waters of Sunderbans including salinity, the most important parameter for the mangrove ecosystem have been changing for several reasons. Almost all the rivers of Sunderbans have lost their connections with the river Hooghly due to the change of the river course for the gradual easterly tilt as a result of neo-tectonic movement (Das 2009a, b) and for that reason fresh water supply to these estuarine rivers of Sunderbans have been stopped. The Sunderbans river waters are almost saline along the entire stretch of the rivers due to lack of head water discharge. Only the river Hooghly in the extreme west of the Sunderbans meets the Bay of Bengal carrying fresh water from the upstream. As a result, tidal sea waters coming through the estuarine river mouths of 2–8 km width to the upstream have become more saline. Sewage discharges, urban emissions, accidental spillage of toxic pollutants in the river waters are the inputs from anthropogenic interferences. The quality of brackish waters has been gradually deteriorating because of squeezing of areas for rapid encroachment of mangrove regions either for the agricultural purposes or for the aquaculture of both prawn and fish.

Some physico-chemical parameters of the river waters of Sunderbans of late have been gradually changing. Salinity around 1 ppt is observed in the pre-monsoon time in the Hooghly river water even at Nurpur which is about 50 km upstream from its confluence to the Bay of Bengal. Salinity of the river waters has been declining very slowly due to sea level rise and encroaches into the fresh water domain. A temperature variation of river water during winter is a result of shallowing of river beds due to rapid rate of siltation. Occasional low dissolved oxygen level is observed either for heavily sediment laden waters or for the mixing of sewage released from the urban areas (Fig. 3.2). All these changing parameters in the river waters of the Sunderbans have vast impacts on the growth and distributions of the mangroves. The expansion of mangroves into the salt marsh areas is observed in the Sunderbans due to increased tidal amplitudes and changes in the water quality of estuarine rivers.

References

Anonymous (1987) A long term multidisciplinary research approach and report on Mangrove ecosystem of Sunderbans. Department of Marine Science, University of Calcutta

Bandyopadhyay BK, Sen HS, Maji B, Yadav JSP (2001) Saline and alkali soils and their management. ISCAR monograph series-1. Port Canning

Brunn P, Nayak BU (1980) Manual on protection and control of coastal erosion in India. NIO, Dona Paula

Das GK (2008) Trend of salinity decline: a case study from the coastal waters of Bakkhali-Frasergunj, India. Indian Sci Cruiser 22(2):9

Das GK (2009a) Beach processes of Tidal Islands of Hugli Estuary, West Bengal. Geogr Rev India 71(3):240–248

Das GK (2009b) Grain size analysis of some beach sands from the Indian coasts. Geogr Rev India 71(1):10–18

Das GK (2012) Impact of water quality on the changing environmental scenario of Sunderbans. Reason. Kalyani Govt. Engineering College XI:57–66

Das GK, Dutta S (2004) Studies on the impact of water quality on the adjoining wetland ecosystem of Bidyadhari River, West Bengal. Indian Sci Cruiser 18(4):16–21

Das GK, Dutta S (2007) Water temperature during winter in the tropical tidal river waters of Sunderbans, India. J Indian Soc Coast Agric Res 25(1):77–78

Das GK, Dutta S, Sanyal SK (2004) Need for Geomorphic mapping in terms of physico-chemical analysis of the sewage fed Bidyadhari River carrying effluents from the greater Calcutta. J Indian Soc Coast Agric Res 22(1&2):49–51

Dodds WK (2009) Eutrophication of US freshwaters: analysis of potential economic damages. Environ Sci Technol 43:12–19

IPCC (2007) Climate change – 2007. Impacts, adaptations and vulnerability. Contribution of Working Group II to the fourth assessment report of the Intergovernmental Panel on Climate Change. Cambridge University Press, Cambridge

Medina E, Lugo AE, Novelo A (1995) Mineral content of foliar tissues of mangroves species in Laguna de Soutecomapan (Veracruz, Mexico) and its relation to salinity. Biotropica 27(3):317–323

Naido G (1990) Effects of nitrate, ammonium and salinity on growth of the mangrove *Brugeira gymnorrhiza* (L) Lam. Aquat Bot 38(2–3):209–219

Ridd PV, Sam R (1996) Profiling ground water silt concentration in mangrove swamps and tropical salt flats. Estuar Coast Shelf Sci 43(5):627–635

Selvam V, Mohan R, Ramasubramaniam R, Azariah K (1991) Plant communities and soil properties of three mangrove stands of Madras coast. Indian J Mar Sci 20:67–69

Chapter 4
Substrate Soils

Abstract The humic and acid sulphate soils of Sunderbans are basically saline micaceous deltaic alluvium which is ideal for the growth of luxuriant mangrove forest. Substrate soils have a major impact on mangrove nutrition and growth. High nutrient concentration and low salinity of substrate soils produce rapid growth of mangroves in Sunderbans. On the basis of salt concentrations present, soils of Sunderbans are classified into saline soil, saline–alkali soil, non-saline alkali soil and degraded saline-alkali soils. Most of the mangrove species prefer low saline substratum. The water holding capacity and process of aeration are poor in the substrate soils of different geomorphic units of Sunderbans.

Keywords Sunderbans • Substrate soil • Soil salinity • Alkalinity • pH • Organic carbon • Nutrients • Mangroves

Sunderbans soil is typically a physiologically dry soil because plants cannot absorb water properly from the soil due to presence of high amount of salt. The coastal tidal forest, Sunderbans has the saline micaceous deltaic alluvium of humid region. Sunderbans has too saline humic and acid sulphate soils of humid tropical region. Acid sulphate soil has highly acidic pH, high EC, presence of humic (organic) horizon, dominance of sulphate and chloride salts. Substrate soils of different geomorphic regions like riverbank, mudflats and natural levees of Sunderbans are poor in aeration and water holding capacity.

The substrate soils of Sunderbans are ideal for the luxuriant growth of mangroves-the tidal forest of coastal region. Soils are usually deposited in the convex margin of rivers, meandering creeks and tidal inlets of the estuarine Sunderbans (Blasco 1977). Soils help in the progradation of coastal Sunderbans facing the Bay of Bengal. The ideal mangroves soils are found in the low lying alluvial coastal plains, drowned valleys, estuarine and deltas carrying water rich in suspended matter with least disturbed marine waters (Vannucci 1989).

Mangroves of Sunderbans are physiologically tolerant of high salt levels and have mechanisms to obtain freshwater despite the strong osmotic potential of sediments (Ball 1996). During absorption of water from soil a few mangrove species excludes salts with ultra filters present in their root systems. Some mangrove species store salt in their leaves and excrete them through specialized salt glands during seasonal

G.K. Das, *Estuarine Morphodynamics of the Sunderbans*,
Coastal Research Library 11, DOI 10.1007/978-3-319-11343-2_4

shedding. These species avoid salts through such salt excretion and allow more salt into the xylem. Rest of the mangrove species store salt in leaf vacuoles and convert into succulent. Sometimes some mangrove accumulates salt in the bark or the wood transferring salts from the leaves.

4.1 Soil Formation

The muddy substratum of the islands of Sunderbans generally contains about 12 % sand, 66 % silt and 22 % clay while mud flats of the Hugli – Matla – Roymangal estuarine area contain about 15 % sand, 80 % silt and 5 % clay. Ganga alluvium and its salinised part are considered as parent materials of the Sunderbans soils. Ganga alluvium is normally salt free where calcite or magnesite is present as rich divalent in this soil. Major quantity of this alluvium drifted into the Bay of Bengal, get salinised and partially deposited at the river banks, mudflats, river flood plains or natural levees at estuarine region after returning from the sea in suspension through numerous tidal creeks, channel and river water courses (Fig. 4.1). Due to

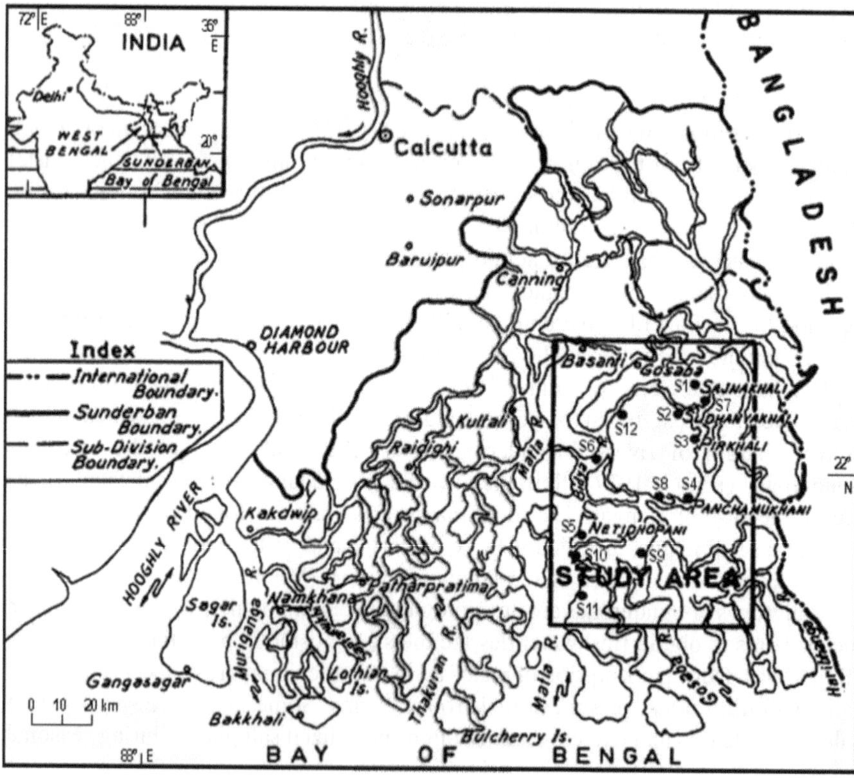

Fig. 4.1 Map of the Sunderbans showing sediment sampling locations

admixture of Sodium Chloride (NaCl) present in seawater, silt and clay of Ganga alluvium undergo partial transformation in their exchange complex for the exchange reaction. There are mainly two soil-forming agencies – (i) rain water and (ii) sea water. Normal soils are produced due to partial leaching by rainwater in presence of calcite, magnesite or dolomite. Saline alkaline soils are produced due to leaching by seawater in presence of excess NaCl but lacking dolomite. Saline alkaline soils become non-saline alkaline soil when salts get completely leached away and sodium ion enters the exchange complex. Degraded alkali soils are produced when hydrogen ion from organic acids takes part in formation of these soils (Das 2011).

4.2 Soil Characteristics

Salt affected soils are classified on the basis of salt concentration measured as Electrical conductivity (EC) and on Exchangeable Sodium Percentage (ESP) of soil. Bandyopadhyay et al. 2001 describes EC as the reciprocal of electrical resistance and has a unit reciprocal of Ohoms i.e. mhos or Siemens (S) in S.I. units 100 S is equal to 1 m mhos unit. The electrical resistivity is the resistance in ohms of a conductor, metallic or electrolytic which is 1 cm long and has a cross sectional area of 1 cm^2. Electrical conductivity (EC) is, therefore, expressed in reciprocal ohms per cm or mhos per cm.

The dividing line between a saline and a non-saline soil was established in terms of electrical conductivity of saturation water extract of soil (ECe) of 4 d Sm^{-1} at 25 °C (Bandyopadhyay et al. 2001). Osmotic potential (Ψ) can be estimated from the electrical conductivity of soil solution or saturation extract of soil (ECe in dSm^{-1}) by the formula, $\Psi = 0.36$ ECe, where, Ψ is the osmotic pressure in atmosphere, ECe is the EC of saturation extract of soil in dSm^{-1}. ECe is also related to total concentration (ppm) of salts by the formula,

$$\text{Salinity (ppm)} = 640 \times \text{EC}\left(\text{dsm}^{-1}\right) \quad \text{or,} \quad \text{Salinity (ppt)} = 0.64 \times \text{EC}\left(\text{dSm}^{-1}\right)$$

In Sunderbans, soils are rich in salts due to saline ground water table present at shallow depth. The estuarine region is generally low lying deltaic plain inundated regularly by brackish waters. Surface soils are salinised due to capillary rise of saline water depending upon the elevation, soil texture, climate, drainage and other reasons (Bandyopadhyay et al. 2001).

4.3 Soil Classifications

Sunderbans soils are classified into four categories namely (i) Saline soil (ii) Saline alkali soil (iii) Non-saline alkali soil and (iv) degraded saline alkali or Saline turf soil.

Types	Characteristics
Saline soil	(i) Inundated
	(ii) Calcite rich
	(iii) Rain water washed
Saline alkali soil	(i) Sea water washed
	(ii) Active delta forming
Non-saline alkali soil	(i) Tidal deposit deficient in Ca
Degraded saline alkali soil	(i) Inundated mangrove forest
	(ii) Land above sea level
	(iii) Area below sea level

4.4 Saline Soils

Saline soils seen in most of the soils are often identified with the white salt encrust on the soil surface. Saline soils have higher concentration of electrolytes (ECe >4 dSm^{-1} at 25 °C) than the normal soils. These soils have pH less than 8.5, ESP less than 15, and preponderance of chlorides and sulphates of sodium, calcium and magnesium. Salinity varies from 2 to 18 ppt in saline soils. It has an average to high clay percentage and immobile Sesquioxides. There is no lime in saline soils as carbonate and it has 11–12 % saturation of monovalents in exchangeable bases.

4.5 Characteristics of the Saline Soils

Availability of mangrove nutrients in Sunderbans soils is affected by higher concentrations of electrolytes in soils. Presence of excess salts in Sunderbans soils interfere with the uptake of nutrient ions by plants. Chloride and sulphates are the chief salt anions. Nitrate may also sometimes, be present in large quantities. Small amount of bicarbonates may be present but soluble carbonates are absent.

4.5.1 Saline Alkali Soils

Salinity of saline alkali soils ranges between 3 and 12 ppt. This soil pH is more than 8.5 (ESP $\geq$15). The ECe is limitless when it is originated from salts capable of alkali hydrolysis, otherwise it should be less than 4 dSm^{-1} at 25 °C. It has high and uniform clay percentage and immobile Sesquioxides. Lime is present in fine form or small concretions. It has 40–56 % saturations of monovalents in exchangeable bases. This type of soil is quite immature for profile development.

Saline alkali soils are of three types in the Sunderbans as following:

(i) Rich in calcite
(ii) Rich in dolomite
(iii) Formation in the inundated areas

This saline alkali soils have three phases of occurrences as –

(i) Rain water washed
(ii) Sea water washed
(iii) Active delta forming phases

4.5.2 *Non-saline Alkali Soils*

Percentage of soil salinity of non-saline alkali soils ranges between 0.1 and 0.9 % and pH varies from 7.1 to 8.2. It has an average to high clay percentage and immobile Sesquioxides. It has no lime as carbonate. This type of soil too is quite immature for the profile development. It has 30–45 % saturation of monovalents to exchangeable bases.

4.5.3 *Degraded Alkali Soils*

Non-saline alkali soils of Sunderbans sometimes may have ESP $\geq$15, yet pH reading in the surface soil may be as low as 6. These are called degraded alkali soils. Soil reaction of degraded saline soils is low where pH ranges between 6 and 6.9. Salinity varies from 6 to 12 ‰. It has average to high clay percentage and has slightly mobile Sesquioxides. No lime is present as carbonate. This type of soil is immature in the profile development. This saline turf soil has 10–17 % saturation of monovalents in exchangeable bases.

Degraded alkali soils occur only in absence of alkaline earth carbonates in soil. The low pH is due to exchangeable hydrogen. The physical properties of soils, however, are dominated by the exchangeable sodium and are typically similar to those observed for alkali soils. This soil has three phases found at – (i) inundated forest; (ii) cultivated areas above sea level and (iii) cultivated area below sea level.

4.5.4 *Acid Sulphate Soils*

This type of soil is found in the majority islands of Sunderbans. Acid sulphate soils are extremely acidic (pH <3.5–4.0) due to oxidation of pyrites and other sulphidic materials accumulated in soil due to reduction of sulphate salts. These soils are formed due to brackish water submergence for long time. Acid sulphate soils are saline, rich in clay, organic matter and exchangeable Al. They are frequently rich in exchangeable Fe and Mn. pH is very low in this type of soil having ESP <15.

4.5.4.1 Characteristics of Acid Sulphate Soils

Acid sulphate soils are highly acidic in nature with or without jarosite mottles, having toxic content of available Fe and soluble Al. These soils are very poor in available P content. They are very low in pH, have usually jarosite mottles of value less than or equal to 6. The soils have high content of soluble S.

Bandyopadhyay et al. (2001) describes that acid sulphate soils develop as a result of the drainage of soils that are rich in pyrites (FeS_2), leading to the oxidation of the compound to produce H_2SO_4.

$$4Fe_2S_2 + 15O_2 + 14H_2O = 4Fe(OH)_3 + 8SO_4^{-2} + 16H^+$$

Oxidation of pyrites may also lead to jarosite [$KFe_3\ (SO_4)_2.(OH)_6$] or natro-jarosite [$NaFe_3\ (SO_4)_2.(OH)_6$] depending upon the condition of reaction. Presence of straw yellow coloured jarosite mottles (value $\leq$2.5 Y and chroma $\leq$6.0) are some typical features of these soils.

Acid sulphate soils develop under the condition where the production of acid due to oxidation of pyrites exceeds the neutralizing capacity of soils and the pH falls below 4.0. Initially, pyrites accumulate in water logged soils that are rich in both organic matter and dissolved sulphates following the reaction:

$$2Fe_2O_3 + 8SO_4^{2-} + 16CH_2O + O_2 \rightarrow 4FeS_2 + 16HCO_3 + 8H_2O$$

4.6 Physico-chemical Properties of Soil

Physico-chemical properties of the soils play very important roles as because a large number of elements are required for biological production.

4.6.1 Soil Reaction (pH)

Soils of the brackish water estuaries of the Sunderbans are generally alkaline in reaction and ranges between 7.05 and 7.96 (Table 4.1). Sometimes it goes down to 3.8 as it found in the anoxic reduced environment. pH of soil is one of the most important factors for maintaining the productivity of any water body since it controls

Table 4.1 Estimation of salinity of sediment samples collected from different locations from Sunderbans

Sl. No.	Sample locations	Sample no.	Salinity (ppt)	pH
1	Sajnekhali (river bank)	S_1	5.76	7.05
2	Sudhanyakhali (river bank)	S_2	5.60	7.19
3	Panchamukhani (river bank)	S_3	4.95	7.05
4	Pirkhali (river bank)	S_4	5.79	7.25
5	Neti Dhopani (river bank)	S_5	5.80	7.35
6	Bidya (river bank)	S_6	4.93	7.3
7	Paschim Sripatinagar (mud flat)	K_5	4.09	7.30
8	”	K_6	5.36	7.68
9	”	K_7	5.04	7.27
10	”	K_8	3.44	7.75
11	”	K_9	2.86	7.35
12	”	K_{11}	3.44	7.05
13	”	K_{12}	2.86	7.69
14	”	K_{13}	2.86	7.19
15	”	K_{14}	3.48	7.73
16	”	K_{15}	3.16	7.75
17	Upendranagar (mudflat)	L_1	3.69	7.18
18	”	L_2	4.81	7.31
19	”	L_3	5.18	7.96
20	”	L_5	4.78	7.56
21	”	L_6	3.92	7.37
22	”	L_8	2.88	7.63
23	Achintyanagar (river bank)	P_1	6.4	7.36
24	”	P_2	5.12	7.37
25	”	P_3	3.84	7.83
26	”	P_4	4.48	7.28
27	Banashyamnagar (river bank)	MB_1	5.12	7.41
28	”	MB_2	1.60	7.25
29	”	MB_3	4.48	7.39
30	”	MB_5	5.44	7.28
31	Maipith (river bank)	$T1_5$	3.84	7.83
32	”	T_{16}	3.2	7.45
33	”	T_{17}	2.56	7.87
34	Paschim Sripatinagar (mud flat)	K_{16}	8.00	7.27
35	”	K_{17}	5.44	7.09
36	”	K_{18}	4.80	7.59
37	”	K_{19}	4.48	7.61
38	”	K_{20}	9.60	7.65
39	”	K_{21}	4.16	7.34
40	”	K_{22}	6.72	7.24

most of the chemical reactions. The availability of nutrient elements through mixing, rate of mineralization of organic matter, fixation of P and other elements and growth and survival of different biotic communities are greatly influenced by pH. Soils are classified into acidic (pH <7.0), neutral (pH = 7.0) and alkaline (pH >7.0) according to its pH or H^+ ion concentration.

4.6.2 Organic Carbon

Formation of organic carbon in the estuaries is slow due to presence of salinity in the river waters of the Sunderbans. Organic carbon acts as a source of energy for the microbes participating in the various biochemical processes resulting release of different nutrients. Soil organic matters or humus influence the physical, chemical and biological activities in soil, improves soil structure, aeration, increase water holding capacity, buffering and exchange capacity of soil including solubility of soil minerals and serves as a store house of various nutrients essential for biotic productions. The observed high organic carbon content is perhaps due to difference in texture, river discharge carrying a large amount of humus and high organic production in the surface waters. Organic carbon and total organic matter can contribute to the variations of the different physico-chemical parameters and have significant positive correlation with several trace metals like iron and copper and their oxides. Organic carbon of sediments varies between 0.04 and 1.20 % (Table 4.2) in the tidal mudflats and river flood plain of Sunderbans.

Table 4.2 Physico-chemical parameters of the mangrove sediments

Sl. No.	Seasons	pH	Salinity (‰)	Organic carbon (%)	Total organic matter (%)
1	Pre-monsoon	7.30	4.09	0.22	0.38
2		7.68	5.36	0.32	0.55
3		7.27	5.04	0.41	0.73
4		6.75	3.44	0.39	0.66
5		7.35	3.86	1.20	2.06
6	Monsoon	7.05	3.44	0.38	0.67
7		7.69	2.86	0.81	1.38
8		7.19	2.86	0.33	0.57
9		6.73	3.44	0.73	1.26
10		6.75	3.16	0.23	0.40
11	Post-monsoon	7.18	3.69	0.35	0.60
12		7.31	4.81	0.29	0.50
13		6.96	5.18	0.14	0.24
14		7.56	4.78	0.30	0.52
15		7.37	3.92	0.04	0.07

4.7 Soil Nutrients

Soil nutrients enrich the flora particularly the world famous mangroves in the ecosystem of the Sunderbans. Among these nitrogen, phosphorus and potassium are termed as primary nutrient elements, calcium, magnesium and sulphur are termed as secondary nutrient elements and boron, copper, manganese, zinc, molybdenum, iron and chlorine are termed as micronutrients on the basis of requirement.

4.7.1 Nitrogen

In the Sunderbans soils nitrogen occurs mostly in organic combinations and inorganic nitrogenous compounds (NH_4, NO_3 and NO_2) are released through bacterial decomposition of organic matter. It is easily decomposable form of organic nitrogen known as available nitrogen, which is important in the aquatic productivity. Nitrogen stimulates primary production in the aquatic environments as a basic and primary constituent of the protein and is essential for the formation of living matter.

Most of the saline soils are poor in available and total nitrogen. Efficiency of plant roots to absorb nutrients from saline soils is also poor due to reduced root volume and various physiological reasons. Nitrogen is present in such soil mostly in organic form and all the inorganic forms of N comprise less than 2 % of total nitrogen in soil. In saline soil, NO_2^- and NO_3^- are present mostly in diffusible forms, whereas NH_4^+ is present mostly in exchangeable form. The process of diffusion is slow in alkali soil condition. The rate of organic matter mineralization is quite slow in the saline soils due to high pH and salinity. Rate of nitrification is too slow in this soil because of high salinity and alkalinity.

4.7.2 Phosphorus

In highly acidic acid sulphate soils of Sunderbans, phosphorus deficiency becomes one of the limiting factors for growth of mangroves species and its associated plants. At high soil salinity, the uptake of P by the mangroves may also be severally affected due to competitive inhibition caused by Cl^- since both phosphates and chlorides, being anions, are absorbed by essentially the same absorption sites.

Phosphorus is essential for assimilation of nitrogen into cellular matter besides respiration, cell division, metabolism, growth and synthesis of protein. Thus it is considered a key element in biological production in aquatic environment and is very often become a limiting factor in fauna food organisms' production. The in situ phosphorus status of most soils is rather low compared to nitrogen and potassium. Phosphorus is absorbed by the mangroves mainly in ionic forms of $H_2PO_4^-$ and

$H_2PO_4^{2-}$. In saline soils, the association of P with other elements depends upon soil pH. So that the chemistry of phosphorus is more complex in saline soils compared to normal soils.

4.7.3 *Potassium*

Potassium helps in the formation of protein, chlorophyll and in stimulating the growth of aquatic plants. Compared to nitrogen and phosphorus, the importance of potassium in aquatic production is less recognised due to its low requirement and easy availability in the Sunderbans soils. Predominance of micaceous minerals is one of the reasons for the high content of K in the saline soils of the Sunderbans. Sodium – potassium exchange in biotite and dissolution of muscovite structural units release large amount of K in alkaline sodium environments. The mica and feldspar (orthoclase and microcline) constitute the major K bearing minerals in saline soils. Clay minerals also constitute an important source of soil potassium. In saline soils, the exchangeable cations decrease the K fixation in the order Na > Ca > Mg > H > NH_4. Saline soils undergo less leaching losses of K because of impeded drainage condition. Growth of the mangroves is accelerated by potassium. It helps in the formation of chlorophyll and protein of the coastal vegetations. The importance of potassium in this aquatic environment is comparatively less due to its low requirement and easy availability in the mangrove sediments compared to that of nitrogen and phosphorus.

4.7.4 *NPK of Sunderbans*

Chemical analysis of sediment samples reveals that the sediments are enriched with nutrients. Estimation of NPK values revealed that available nitrogen (Av. N) ranges from 880 to 1,540 kg ha^{-1}, available phosphorus (Av. P) from 11.4 to 26.0 kg ha^{-1} and available potash (Av. k) from 2,083 to 2,958 kg ha^{-1}. The organic carbon shows a range between 0.38 and 0.69 % and the total organic matter in sediments 0.066–1.19 % (Table 4.3). It is found that available potash is five times richer than that of required amount for the natural growth of mangroves, which supply the nutrients to the ambient environment (Das 2010). Nitrogen to phosphorus ratios indicate that the major portion of phosphorus is of abiogenic origin.

4.8 Micronutrients

Calcium is an integral part of the plant tissues. Sulphur is an essential constituent of protoplasm. The concentration of CO_2 in water is influenced by Ca and Mg. Calcium also acts to increase the availability of other ions in water and in general ameliorates

Table 4.3 Chemical analysis of collected sediment samples of deltaic Sunderbans

Samples no.	pH (1:2)	Salinity (ppt)	Av. N (kg/ha)	Av. P (kg/ha)	Av. K (kg/ha)	Organic carbon (%)	Total organic matter (%)
S_1	8.05	5.67	1,100	18.6	2,285	0.47	0.81
S_2	8.1	5.6	1,540	17.8	2,554	0.69	1.19
S_3	8.0	4.95	1,540	26.0	2,083	0.67	1.16
S_4	8.25	5.79	1,320	26.0	2,957	0.63	1.09
S_5	8.25	5.8	880	20.5	2,419	0.38	0.66
S_6	8.3	4.93	1,100	11.4	2,688	0.47	0.81

the chemical conditions of water. The most important factors influencing the availability of micronutrients in saline soils are pH, soil texture, type and quality of clay minerals, organic matter, calcareousness and salt content in soil. In almost all types of saline soils, there is a negative correlation between soil pH and available micronutrient cations. A highly significant negative correlation has been reported by Bandyopadhyay et al. (2001) between extractable Zn, pH and $CaCO_3$ content of soil.

Calcium is generally present in soil as calcium carbonate. The amount of exchangeable phosphorus in the sediments is inversely related to the calcium carbonate – organic matter ratio so that in highly organic soil with low calcium soluble phosphorus remain adsorbed in exchangeable forms and when sediment is very low in organic matter and high in calcium, phosphorus is fixed as insoluble precipitate. Some micronutrients are essential in the mangrove ecosystems and the brackish waters contain considerable amount of Ca, Mg, S and micronutrient elements.

The deficiency of Fe, Mn and Cu is quite frequent in saline alkaline soils particularly when the soil is calcareous in nature. Many a time boron is accumulated in salt affected soils in toxic concentrations due to poor leaching. The native source of this boron in coastal soils is the primary mineral tourmaline. Saline soils sometimes show deficiency of Mo and it is seen that in saline soils Mo availability increases with pH. Zn has the specific role in improving the alkali tolerance of mangroves.

It is observed that in general, saline soils have better availability of micronutrients than the saline alkaline soils in Sunderbans. Highly acidic saline soils or acid sulphate soils of Sunderbans contain toxic concentration of Fe and Mn, particularly in the inundated condition. Al toxicity is seen in acid sulphate soils.

4.9 Metal Content and Metal Oxides of Soils

Sunderbans river estuaries are unique system as it shelter and support various living and non-living resources. In this system some elements of biologically active materials and a few organic compounds play an important role for the productivity of this environment. Some metals and non metals derived from several sources, are carried by the river water and spend considerable period where they are subjected to the influence of tidal flux and experience various bio-geo-chemical transformations. Nitrogen, phosphorus, potassium, carbon, trace metals and organic matter studied in the estuarine sediments of Sunderbans covering a large undisturbed part of the sanctuary areas, provide the clear picture of their occurrence, distribution with reference to productivity and to identify their probable sources.

Collected soil samples are analysed with respect to their metal oxides and their values are furnished in the Tables 4.4 and 4.5. The oxides are shown in percentages while metal contents are shown in parts per million (ppm). Amongst the heavy metals, copper varies from a minimum of less than 10 ppm to a maximum of 100 ppm, while lead and zinc varies from less than 10 ppm to 40 ppm and from

Table 4.4 Metal oxides of the sediment samples collected from the Sunderbans

Sample no.	Fe_2O_3 %	MgO %	Na_2O %	K_2O %	L.O.I %	ROOM % as 'C'
S 1	6.27	1.86	0.91	2.84	8.94	0.62
S 2	5.48	1.67	0.82	2.40	7.86	0.59
S 3	5.77	1.84	1.00	2.93	8.64	0.61
S 4	5.80	1.80	1.14	2.52	8.25	0.47
S 5	5.56	1.70	1.02	2.39	8.25	0.61
S 6	5.14	1.57	1.04	2.41	7.15	0.64
S 7	6.12	1.66	0.83	2.45	8.65	0.55
S 8	5.98	1.81	0.95	2.30	8.21	0.61
S 9	5.72	1.63	0.93	2.67	8.72	0.57
S 10	6.24	1.47	1.01	2.59	8.72	0.47
S 11	5.88	1.32	0.91	2.64	8.56	0.51
S 12	5.83	1.53	0.89	2.77	8.82	0.58

Table 4.5 Analysis of metal contents (ppm) in the sediments of Sunderbans

Sample no.	Cu	Pb	Zn	Ni	Co	Cd	Cr	Mn
S 1	70	70	80	40	30	<5	65	620
S 2	40	60	60	30	30	<5	65	560
S 3	45	60	65	35	35	<5	70	620
S 4	40	65	60	45	30	<5	55	580
S 5	40	65	60	35	30	<5	55	545
S 6	35	55	55	30	30	<5	50	520
S 7	60	55	70	40	35	<5	65	570
S 8	55	60	65	30	30	<5	70	530
S 9	65	65	60	35	30	<5	60	610
S 10	45	55	45	40	35	<5	55	540
S 11	55	45	40	45	30	<5	60	580
S 12	65	40	75	35	35	<5	50	555

18 ppm to 30 ppm respectively. Generally the metal content is less in the clayey sediments and the higher values are restricted in the finer fractions like sandy silts and/or silt. Mangroves themselves, however, generally have low concentrations of heavy metals and they are very poor indicators of trace metal concentrations. Among metals, Zn showed low value in almost all species of plants and animals followed by Cu and Pb because the low level of metals in the mangroves themselves may be due to low bio-availability in the mangal sediments (Table 4.5). Disturbances like changes in the frequency and duration of tidal flooding; changes in salinity or prolonged dry periods cause the mangrove soils to lose their metal binding capacity, resulting in mobilization of the metals. These disruptions are often associated with anthropogenic stresses (Lacerda 1998). The seasonal variations of trace metals and metal oxides are probably due to the difference in existing physico- chemical condition and the adjacent land effect of the estuarine region.

4.10 Soil Distribution

Saline soils are identified in the region of Sandeshkhali II, Hasnabad, Hingalganj, Gosaba, Kultali, Jaynagar II and Mathurapur II. Soils are saline alkaline at Kakdwip, Kulpi and Sagar blocks. Non-saline alkali soils are found at Mathurapur block and degraded saline alkali soil is found in Ghusighata, Sandeshkhali, Tushkhali and Damakhali regions (Anonymous 1987).

4.10.1 Saline Soils

Entire region of Hingolganj and Hasnabad blocks lying west of Sahebkhali River are saline. The northern part of Joynagar II, Kultali and Mathurapur II are saline. A small area lying south of Bidya river in Gosaba and Sandeshkhali II is also saline.

Hatkhali and Bidya River. They occur in southern portion of Gosaba and Hingalganj.

4.10.2 Saline: Alkaline Soils

The region of Mathurapur II having Mani river, Saralda khal, Raidighi river, Kaloa K, Jalaberia K, Chuprijhara R, Chatu R and Thakuran river on its south and small streams on the north have saline – alkali soils. In Joynagar II, east and south part of Kulpi block show this type of soils. A considerable area in Sagar Island and Kakdwip blocks is under such soils mainly south of Kalnagini K and Ghugudanga R.

4.10.3 Non-saline Alkali Soils

This type of soils is found in a part of Mathurapur II block only in tidal deposits deficient in calcium.

4.10.4 Degraded Saline Alkali Soils

This soil is found in Sandeshkhali block lying between Hingalgunj and Haroa blocks.

4.11 Summary

The high salt content and/or high pH in different types of saline soils cause changes in solubility, availability and efficiency of nutrients for the growth of mangroves in Sunderbans. Nitrogen uptake by a few mangroves grown in the relatively upper middle portions of the islands is affected by the presence of high concentration of Cl^- ion. Further high sodium can cause calcium and magnesium deficiencies. Most of the mangrove species prefer low saline substratum. The mangrove species inundated by two times daily of tides prefer soils hypersalinity areas whereas species those occur in the central portion of islands prefer their habitat with low salinity. Mangrove species diversity and richness depend upon the preferred salinity of respective mangrove species. Mangrove zonation is happened in response to the salinity of both soil and water. Species diversity of mangroves is controlled not only by the salinity alone but by the several factors like availability of nutrients (NPK), organic carbon, soil texture, water holding capacity of soil etc. Substrate soils have a major impact on mangrove nutrition and growth. High salinity and alkalinity of the soils of Sunderbans delay nitrification rates and urea hydrolysis due to poor bacterial activities in soil. The process of diffusion is slow in alkali soil condition. Both saline and alkali are usually low in available nitrogen and organic matter, and have poor contribution from symbiotic and non-symbiotic nitrogen fixation. High nutrient concentration and low salinity of substrate soils produce rapid growth of mangroves in Sunderbans. Clearing of mangroves forest or formation of canopy gaps can change the physical and chemical characteristics of the underlying soils, leading to the anaerobiosis and increased sulfide in the sediments. Mangroves are gradually becoming dwarf due to scarcity of nutrient supply as islands-tops are not inundated on regular basis. Growth of some mangrove species (those prefer less salinity) is reduced because of physiological water deficit on account of high osmotic potential at the root zone soil solution, although there may be sufficient amount of water in soils (Das 2013). Decreasing osmotic potential of soil solution has the net effect of reducing availability of water to plants, although some tolerant mangrove species has the ability to adjust its internal osmotic potential by various means like through production of organic acids, uptake of salts etc. (Bandyopadhyay et al. 2001). Mangroves grown in the islands of late show stunted growth, smaller and thicker leaves, and dark green or bluish green in colour due to decreased cell elongation and their leaf surface is layered with waxy materials or scorching of leaf margin.

References

Anonymous (1987) A long term multidisciplinary research approach and report on Mangrove ecosystem of Sunderbans. Department of Marine Science, University of Calcutta, Kolkata

Ball MC (1996) Comparative eco-physiology of mangrove forest and tropical low land moist forest. In: Mulkey SS, Chazdon RL, Smith AO (eds) Tropical forest plant ecophysiology. Chapman & Hall, New York, pp 461–469

Bandyopadhyay BK, Sen HS, Maji B, Yadav JSP (2001) Saline and alkali soils and their management. ISCAR monograph series-1, Port Canning

Blasco F (1977) Outline of the ecology, botany and forestry of the Mangals of the Indian subcontinent. In: Chapman VJ (ed) Ecosystem of the world. Elsevier, New York, pp 241–262

Das GK (2010) Biodiversity in Sunderbans. In: Dwivedi AK, Tripathi SC (eds) Environment and biodiversity, environmental pollution and risk to biodiversity. Lambert Academic Publishing AG & Co., Saarbrucken, pp 1–34

Das GK (2011) Impact of salinity and nutrients on the changing mangrove floristic- a case study from the River Flood Plains of Sunderbans, India. In: Biotic potential and the abiotic stress. Lambert Academic Publishing AG & Co., Saarbrucken, pp 119–129

Das GK (2013) Characteristics of mangrove substrate sediments of Sunderbans. Reason. Kalyani Govt. Engineering College, Kalyani XII:7–18

Lacerda LD (1998) Trace metals biogeochemistry and diffuse pollution in mangrove ecosystems. ISME Mangrove Ecosyst Occas Pap 2:1–61

Vannucci M (1989) The mangroves and us; a synthesis of insights. Indian Association for the Advancement of Science, New Delhi

Chapter 5
Coastal Morpho-Dynamics

Abstract Coastal Sunderbans lies in the southern part of the Ganga – Brahmaputra delta which is the largest delta system of the world. Morphodynamic of Sunderbans coast is a combination of various strong dynamic processes which bring discernable changes in the coastal configuration within short span. Ocean current and waves, strong wind and cyclone surges, active hydrodynamics and neotectonics, acting singularly or in tandem, are main agents acting in these changing processes. For the interpretation of coastal morphodynamics and delineation of specific environmental zones, field studies of various bedforms occurring in the area, mud balls etc. are important. Bakkhali beach provides an example of dynamic morphological domain which is presently under threat of severe erosion. Flow regime change is recorded with the change of bedforms from lower foreshore zone to upper foreshore zone.

Keywords Morphodynamics • Storms • Tides • Coastal dunes • Beach • Bedforms • Bakkhali • Sagar Island • Flow regime

Reconstruction of the earth history depends on the interpretation of ancient sedimentary environments, which, in turn, depends in part, on studies on sedimentary structures of both bed forms and their related stratifications. Migrating bed forms are often responsible to create internal bedding structures particularly the various forms of cross bedding. There are several works by McKee (1957a) and Harms and Fahenstock (1965) – showing relation of bed forms and bedding structures. The surface structures profusely occurring bed forms, mud balls etc. are quite sufficient to differentiate the specific environmental zones of the intertidal beach-face and interpretation of the coastal morpho-dynamics.

5.1 Coastal Environment

The Bay of Bengal holds the Bakkhali and Sagar Island, the geological marvels on the southern part of Sunderbans of Bengal deltaic plain in the Bengal Basin. The century old sea beach of Bakkhali and Gangasagar known for recovering of health and spending leisure at the vacations is quiet peaceful sandy shore with strands of

G.K. Das, *Estuarine Morphodynamics of the Sunderbans*,
Coastal Research Library 11, DOI 10.1007/978-3-319-11343-2_5

casuarinas even upon the dunes bordering coastline amidst the vibrant biodiversity upon the beach where the molluscan shells, mud lumps and mud balls are scattered. With a coastal length of about 1.2 Km Bakkhali (Lat 21°34′ N, Long 88°16′ E) is situated in the lower reaches of the South 24-Parganas district along the coastal tract of West Bengal, India. This sea beach keeps changing their beach characters after every monsoon.

5.2 Sediment Transport and Shoreline Progradation

The total sediment load of the Ganga-Brahmaputra River is long known to be much higher than any other river system of the world (Coleman 1969; Milliman and Meade 1983). Nevertheless, there has been a long debate about the state of the progradation of the delta as well as the shoreline since the last 200 years. According to Morgan and Mclwtire (1959) and Coleman (1969) very insignificant delta progradation has occurred during this period. The following opinions as regarding this steady state of the shoreline (Allison 1998) have been put forward: (i) the huge alluvial sediments instead of being added to delta progradation are trapped by a subsiding trough that existed along the axis of the Ganga-Brahmaputra rivers, (ii) rapid subsidence of the basin blocked the optimum development of delta margin. Intermittent structural activities stood in the way of systematic progradational growth of the delta and (iii) the bulk of sediments derived from the upland regions rather by-passed to contribute to the Bengal deep sea fan through the sub-marine canyon “Swatch of No ground”.

However, Roy Barman (1983, 1992) suggested that the rate of sediment supply from the Himalayas and North Eastern Hills was more than the rate of subsidence of the Bengal basin. As a result, the basin of sedimentation filled up rapidly and the Bengal delta started prograding southward. According to him, the role of tectonism was surprisingly small in the deep basin and shelf areas of West Bengal with the only exception of some down-to-basin faults in the basin margin and shelf-slope break regions. It is only recently from high-resolution seismic survey Kuehl et al. (1997) have established that the Holocene sub-aqueous delta of the Ganga is a prograding one with a characteristic clinoform stratigraphy with bottomset, foreset and topset strata in lateral progradation. Allison (1998) has also put forward some evidence for recent growth of this delta in its sub-aerial portion together with a lateral progradation.

5.3 Bakkhali and Fraserganj Coastal Environments

Bakkhali beach stands on the east of the river Hugli along the vast stretch of about 180 km coastal tract of West Bengal. This sea beach represents a wide range of diversity in terms of coastal processes, geomorphology, environment etc.

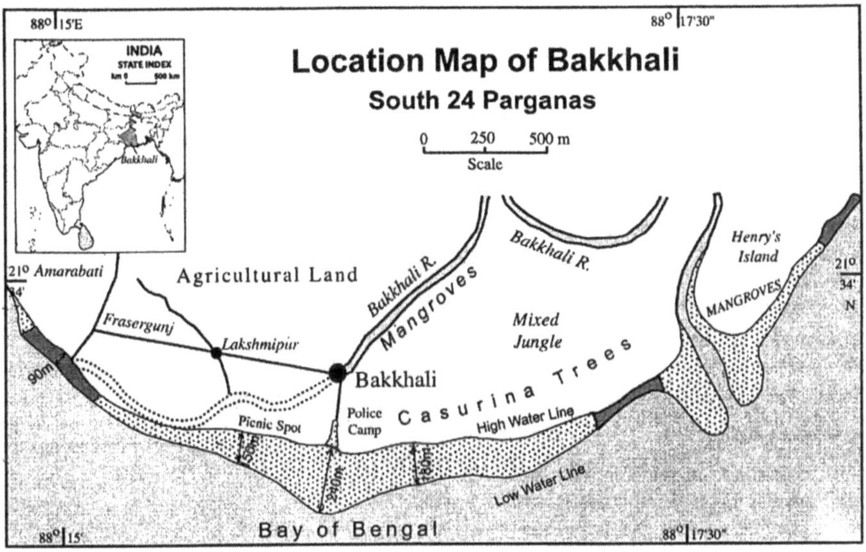

Fig. 5.1 Location map of Bakkhali

West Bengal coast of India is characterized by the presence of the largest tide dominated Hugli estuary with numerous channels, creeks and impressive digitize architecture, supposed by the extensive mangroves (Das 2009a, 2010). Bakkhali sea beach along the West Bengal coast (Fig. 5.1) stand on the lower (tidal) delta plain of the Ganges-Brahmaputra delta. This coastal stretch was lying beyond the paleo-shoreline during Holocene time in terms of available evidences indicating that maximum sea level transgression occurred at about 6,500 BP when the shoreline was low to 300 Km inland of the present shoreline. Shoreline progradation and basin infilling by the Ganges-Brahmaputra rivers subsequent to this time accounts for 30,000 km^2 of growth comprises 30 % of the modern delta plain. But some of the important factors are still unrecognized as to what extent sediment is supplied to the coastal West Bengal from the river Ganges and its distributaries or from marine side (Bhattacharya 1988); the process and nature of river sediment supply to the lowland flood plain, lower delta plain and marine end member.

5.3.1 Beach Morphology

The present work is an attempt to study the various bed forms and the related flow phenomena active in the intertidal zones of the tidal islands of the Hooghly estuary, West Bengal. The Gangasagar beach, south of Sagar Island and the beach encircling the Lower Long Sand – an offshore tidal island in the Bay of Bengal, 16 km south of Gangasagar beach have been studied for the purpose. The bed forms occurring in these generally sandy intertidal zones have been examined with special reference to their (i) association and distribution in space, (ii) relation to flow phenomena and (iii) identification of the different intertidal zones from low to high water level based on their association.

The fluctuation of tides in the form of rise and fall of water level causes migration of energy zones (De Boer et al. 1989), which results characteristic surface sedimentary structures dependent on factors like beach slope, local morphology, depth of water, flow direction, flow velocity and textural attributes of sediments. Apropos of the variations of these elements on the beach surface, the surface sedimentary structures show wide variations in their morphological forms and dimensions within small locales of intertidal zones. Lateral variations of these structures as well as their variations at right angles to the coastline have been studied to interpret the hydrodynamic conditions of sedimentation. The upper foreshore zone is characterized by the occurrence of sinuous trains of backwash ripple marks, swash marks and rhomboid marks. The middle foreshore zone with runnels and ridges show assemblages of lingoid ripple marks of large and small scale, interference ripple marks, particularly of polygonal and ladder back type, rill marks, rhomboid marks and water marks on lee and stoss slopes of large scale ripple marks. The lower foreshore zone, however, is characterized by the capped-off ripple marks, ripple- rill complex, double-crested ripple marks and current crescents. The ripple marks occurring along a beach profile from the lower to the upper foreshore zone generally show an increase in the strength of flow belonging to the lower flow regime. Reconstruction of the earth history depends on the interpretation of ancient sedimentary environments, which, in turn, depends in part, on studies on sedimentary structures of both bed forms and their related stratifications. Migrating bed forms are often responsible to create internal bedding structures particularly the various forms of cross bedding. There are several works by McKee (1957b) and Harms and Fahenstock (1965) – showing relation of bed forms and bedding structures. In the present study observations have been confined within the surface structures, as these profusely occurring bed forms are quite sufficient to differentiate the specific environmental zones of the intertidal beach-face (Friend et al. 2003).

These bed forms have been studied in the sandy beach, South Sagar Island and the Lower Long Sand – an offshore tidal island in the Bay of Bengal, 16 km South of Gangasagar beach. These bed forms have been examined with special reference to (i) their distribution in the natural settings in the beaches, (ii) their relation to flow-phenomena and (iii) identification of the different beach zones from low to high water level based on their association.

5.3.2 *Climate*

The climate in the Sunderbans coast is characterized by the southwest monsoon system. Seasonal low-pressure areas over the Persian Gulf initiate the southwest monsoon in summer (May–September) with high precipitation rates. Seasonal high-pressure areas over Tibet create the northeast monsoon in winter

(December–February), which is calm and dry. Changing monsoonal winds strongly affect the surface water flow in the Bay of Bengal (Das 2010). In spring, northeast monsoon to the southwest monsoon surface water movement is characterized by a clockwise rotation. In autumn, during transition from southwest to southeast monsoon, surface water mainly flows counter clockwise. Strong tides govern the channel regime and influence the upstream sediment movement and there is almost no headwater discharge during the summer in the mouth of Hooghly. On the contrary, the channel regime is governed by the combined effect of both the headwater discharge and tides and influence the seaward drift of the sediments during the monsoon.

5.3.3 *Tide*

Tides in the Bay of Bengal near Bakkhali primarily are semidiurnal, with maximum current velocities of 3.8 ms^{-1} and moderate wave activities. The funnel-type shape of the Hooghly-Matla estuary causes tides to be highest along the northeastern coast where ranges can be as high as 6 m near the river mouths, but normally vary between 4 m in the eastern channels and 1.5 m in the western channels. The low tidal height varied from 0.74 to 4.17 m and high tide from 1.67 to 5.46 m along this coast (Das 2009b, 2010).

5.4 Dunes

Dunes are the most spectacular structures of the sea facing delta. The coastal dunes of the Sunderbans looked like dimensionless mound shaped, are transverse and barkhan types having sinuous crests and run parallel to the coast line. They are formed depending upon the wind flow and direction, and have particular association of plants. The formation of the dunes depends on the strong wind actions for the movement of the foreshore sands to the supratidal backshore zone. Three important parameters are responsible for the formation of dunes (Bhattacharya 2000) such as –

(i) Vigour of wind flow and pattern
(ii) Rainfall upon the dunes
(iii) Floral community

The maximum wind velocity of coastal region like Bakkhali- Fraserganj and Gangasagar and river mouth islands like Jambu dwip and Chuksar Island, ranges between 35 and 55 km in summer and 2 and 4 km during post-monsoon time. The direction and wind velocity is slightly affected depending upon the physiography of the coast facing islands and sporadic occurrences of vegetations. The dominant wind direction is towards south and south – south west during pre-monsoon period and towards north to north west during winter seasons. The lee slope of the coastal dunes is observed in the direction of south to south west as per the direction of the blown wind during that period.

The average rainfall along this coastal stretch is 1,953 mm per annum as recorded in the Gangasagar station by the IMD, Kolkata and ranges between 1,900 and 2,100 mm. High moisture contents inside the dunes is obtained during the monsoon times and as a consequence, vegetations grown up over the dunes favoured by this high moisture condition. The grown up plants covering these coastal dunes act as the sand binders that accelerate the morphological growth of dunes.

A few shrubs and creepers are seen to grow over the dunes. Proliferation of these vegetations generally depends upon a few physico-chemical conditions like salinity, moisture content, substrate quality etc. vegetations play important roles like stability of the dunes and arresting of the migration of the dunes. Creepers and shrubs help in the creation of these coastal dunes by arresting sand grains behaving as barriers. The important herbs of the coastal dunes are *Ipomoea pescaprae*, *Salichornia* sp., *Paspalum vaginatum*, *Suaeda maritime*, *Synodon dactylon*, *Aeluropus logopoides*, *Sesuvium portulacastrum* etc. Among all these creepers *Ipomoea pescaprae* play the dominant role for formation, stabilization and growth of the coastal dunes along the coastal tract of Sunderbans. The rootlets of these creepers assist in sand binding and further arrest the migration of such dunes (Van der Maarel 2003).

5.4.1 Roles of Coastal Dunes

Wind pattern and vigour of wind flow, precipitation in the dune areas and plant association play the major role for the formation of sand dunes along the coastal tract of the Sunderbans (Fig. 5.2). These coastal sand dunes form the ultimate border of the shore along the coastline. Vegetation types are related to the formation of dunes and their morphology (Carle and Hill 2009) although the dunes bordering the Bakkhali beach have no definite shapes. The morphology and evolution of the fore dunes are controlled by the abundance of vegetation (Carranza et al. 2010;

Fig. 5.2 Coastal dunes along the coastal stretch of Sunderbans

Vila' et al. 2000, 2006). The dominant herbaceous and grassy vegetations include e.g. *Ipomoea pescaprae* Sweet, *Aeluropus legopoides* Trin. and *Porteresia coarctata* Roxb can stabilize the sand dunes. These plants eventually have manifold roles to play in dune formation as following:

(i) their roots and rootlets act as sediment binders
(ii) the leaves and stems help fall out of suspensional sands
(iii) the plant type help stabilization of dunes
(iv) the dune vegetations arrest dune migration to a considerable extent (Bhattacharya 2000).

It was recorded that about 14 m high sand dune of Bakkhali protecting the coastline was eroded away by cyclonic storm surges in 1985 (Mitra and Samaddar 2005).

5.5 Coastal Processes

Coastal processes are very dynamic and it is induced by the tropical cyclones of the area. Field observations show that the coastline is changing in the beach areas (Fig. 5.3) due to abnormal cliff erosion. The important morphotypes of the beaches are coastal dunes, sand flats, runnels and inlets, mudflats and mangrove patches. Wave parameters, micro- and macro tidal cycle, long shore currents are recorded in the beach areas of Bakkhali.

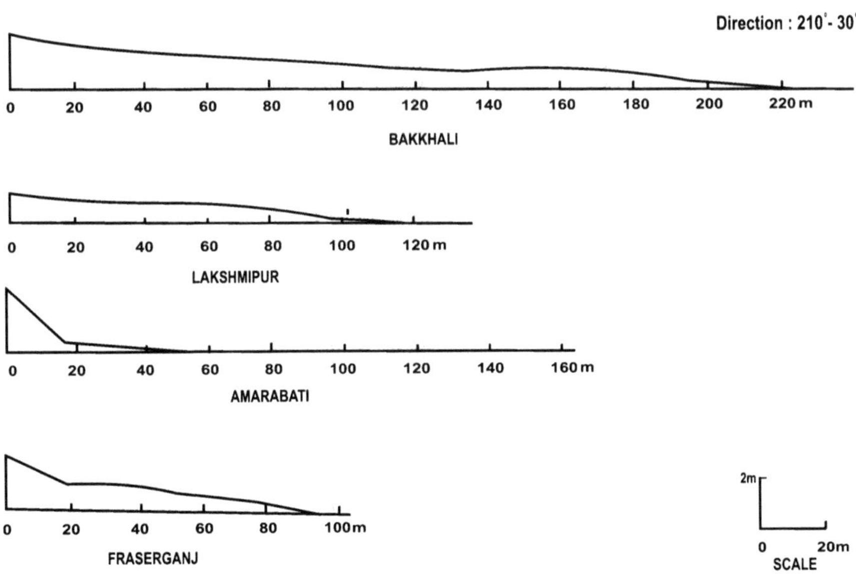

Fig. 5.3 Beach profile of intertidal region at coastal areas at Bakkhali and Fraserganj

Table 5.1 Gastropod and pelecypods species at Bakkhali beach

Gastropods	Pelecypods
Architectonica perspective (Linn)	*Striarca lactea* (Linn)
Cerithidea obtuse (Lamarck)	*Mactra luzonica* (Deshayes)
Natica tigrina (Roeding)	*Apolymetis edentula* (Spengler)
Polinices turridus (Swainson)	*Sanguinolaria acuminate* (Deshayes)
Gyrineum natator (Roeding)	*Meretrix meretrix* (Linn)
Thais lacera (Born)	*Glauconome sculpta* (Sowerby)
Pugilina cochlidium (Linn)	*Barnea candida* (Linn)
Nassarius stolatus (Gmelin)	
Amalda ampla (Gmelin)	

The Bakkhali beach constitutes the interface between the sea and the coastal stretches, shoreline and the coast. The sea waves, modified by the action of tide and wind are the principal architect of morphological changes including erosion in the beaches. The almost 1.2 km stretch of Bakkhali beach from its east wing to the west provides an example of dynamic morphological domain, which is presently under threat of severe erosion. As a result of that the beach width of Fraserganj is very narrow. The severity of erosion is evident from the fact that the underlying clayey sediment is often exposed in upper and middle foreshores of the sandy coast as a result of storm surges. Pebbles, mud balls, fragments of wood and numerous gastropod and pelecypod shells (Table 5.1) occur on the sandy beaches, which obstruct the flow during the backwash and form scour through during strong tidal surge.

5.6 Tidal Shoals

Large sandy shoals are frequently present away from shore up to latitude 21° 30′ N in the south and to the longitude 88° 17′ E in the east of the Bakkhali coast which are only visible at low tide when water recedes from the surface of the sand flat. Sediment deposition occurs on the southern side so that the shoal is accreting towards southeast. The shoals consist of sands with a thin layer of silt-mud laminae on the east. A depressed zone separates the shoal from the beach zone, which is to be considered as ebb channel that shallows in the direction of the ebb. It flows the coastline at a little distance. On the elevated portions of the shoals flood velocity is greater although ebb velocity is usually greater in the depressed zone used as ebb channel. The sediment is coarser on the lower flat area and finer towards upper parts. The shoal is built up predominantly by the tides.

5.6.1 *Bedforms*

These intertidal shoals are characterized with large-scale bed forms like flood and ebb-oriented megaripples superimposed with small-scale bedforms of almost all types of ripple marks. Sand waves found in the eastern part are formed in the strong tidal current situation of 56 cm/s. Although megaripples are common features at the surface of the shoals, sets of ripple laminae, cross bedding, flaser bedding and convolute laminations are developed as the internal laminations of intertidal sandy shoals. Mud drapes occur within the sand layers indicating a transportation of mud from the highly eroded sea beach where mud exposed. Although on sandflats, wavy bedding is common as well as ripples, but laminated sands, flaser bedding and lenticular bedding are more common. Tidal lamination, with sand deposited during ebb and flood and mud deposited from suspension during slack tide have an overall thickness of 4–6 mm, reflecting a full tidal cycle.

5.7 Sediment Texture

The statistical size parameter of beach sample of three different geomorphic zonations of Bakkhali sea beach show that the sediments are generally well-sorted ($\sigma_1 = 0.3510$–0.4515 phi) fine sandy in nature. All the samples of the supratidal zone are slightly negatively to slightly positively skewed ($SK_1 = -0.1221$ to 0.3344 phi) and are mostly platykurtic ($K_G < 1.00$). The cumulative curves are mostly non-linear with the saltation mode of transport as the chief population for distributions (Folk and Ward 1957).

5.8 Composition of Beach Sediments

The heavy mineral of the sediments collected from the beach areas of Bakkhali indicates that the sediments are mostly derived from the crystalline, metamorphic and sedimentary rocks (Das 2004). Zircon, tourmaline, rutile, topaz, barite, hornblende, olivine, apatite, magnetite and ilmenite indicate derivation of sediments from crystalline rocks, whereas epidote, zoisite, garnet, kyanite, sillimanite and biotite are typical mineral of metamorphic origin.

5.9 Micro-paleontology

Biological factors play a significant role along this biogenous coast of Bay of Bengal. Activities of tidal fauna in the intertidal flats of both the beaches develop micro-morphological features, which are very much significant for the trapping of sediments. Foraminifera may be considered as an important environmental indicator for several years. Intertidal beach environment of Bakkhali-Fraserganj is favourable for the prolific growth and occurrence of the benthic foraminifera, *Asterorotalia trispinosa*. Physico-chemical parameters such as salinity 15–31 ppt, pH 7.21–8.37, temp 21–25 °C, D.O. 7.38–8.05 mg/L of the sea, waters of intertidal beach and the sandy bottom with subordinate silt and clay are suitable for the flourishment of foraminifera. *Ammonia* sp., which is found, concentrated at Fraserganj beach sands. It is an environment sensitive taxon and acts as a pollution indicator. Trash pollutes the sea beach that leads to some kind of pollution at the sediment-water interface of Bakkhali. *Ammonia* sp. is almost scarce in this part.

Environmental sensitiveness of *Ammonia* sp. is supported by the observations of low species diversity at the eastern part of the beach. The tolerance level of this taxon against pollution is more with respect to other taxa in the environment. The assemblage of foraminifers like *Elphidium* sp., *Ammonia* sp., *Asterorotalian pulchella*, *Triloculina* sp. and *Parafissurina* sp. indicates shallow water, moderate to low salinity environment of Fraserganj-Bakkhali intertidal beach areas.

On the basis of relative abundance in the intertidal surficial sediments, two benthic foraminiferal biofacies are identified in these dynamic beach areas. *Astrorotalian* sp., occupies the eastern part i.e. Bakkhali side while *Ammonia* sp. concentrates only in the western part i.e. Frasergunj side. Species diversity and the population density of foraminifera are considerably less in the eastern part of the beach area than that of the western part which indicates a higher rate of sedimentation in the former domain i.e. at Bakkhali side than in Frasergunj side. Abundant occurrence of foraminifera by means of population density and species diversity at the western part of Frasergunj side indicates a low rate of sedimentation. Mechanical fragmentation of biota, particularly the benthic foraminiferal assemblage, and their reduced size at Bakkhali sediment suggest a eurihaline shallow marine environment as well as huge influx of freshwater by the Hugli River. High benthic foraminiferal assemblages in comparison to the planktonic foraminifera also indicates nearshore marine dynamic beach environment with high-energy conditions (Rose and Lidz 1977).

5.10 Shoreline Recession

Coastal changes, as a result of Coastal erosion (Fig. 5.4) are measured using the relict sea walls still exposed on the beach during the low tide. This seawall was constructed in order to protect the resorts, bunglows, gardens and other structures during British period (1908–1910). The sea wall was completely destroyed by the devastating cyclonic waves during early 1970s. The present shoreline is extended about 1 km landward from the relict of the seawall. The recession rate of shoreline is calculated in two phases separately. The rate of recession of shoreline at Bakkhali as

Fig. 5.4 Severe erosion undergone on the beach surface at Fraserganj. Wave ripples are seen at the center

recorded by Mitra and Samaddar (2005) is 9.7 m/year during the period 1930–1970 and 5.4–8.5 m/year during 1971–1995. The erosional rate in the west flank of Bakkhali beach is so high that the underlying clayey sediments are exposed in a large area in the intertidal zone particularly during monsoonal period.

5.11 Mud Balls

Mud balls, almost spherical in shape, are scattered in between the supratidal and intertidal areas covering a width of 9 m of the river mouth islands and the coastal Sunderbans facing the Bay of Bengal. Mud balls, derived from the mud chunks that collapsed from the mud banks due to high tidal vigour and wave actions are strewn over about 1.2 km stretch at the transitional zone between upper intertidal and supratidal zone of present day beach of Bakkhali covering a width of 9 m area parallel to the shoreline. These mud balls are almost spherical in shape (Fig. 5.5) and their diameter varies from 3.18 to 7.3 cm. The mud balls attain different shapes like cylindrical, spindle, ellipsoid, sphere etc. before becoming mud balls through the rolling mechanism two times daily in the semidiurnal tidal situations. Molluscan shells either intact or broken (Gastropods and Pelecypods) are found in these mud balls which are either within the mud lumps earlier or inserted later from the beach during rolling by the wave actions. These mud balls, compositionally made up of silty clay are covered with a few fine sand grains and biotic materials added during rolling. Occurrences of these mud balls of mechanical origin throws some light upon erosion of the beach environment that helps in the interpretation of hydrodynamic environment.

Fig. 5.5 Mud ball strewn at the transition zone between supratidal and intertidal region at Fraserganj

5.11.1 Significance of Mud Balls

Analyses of the foraminiferan community found inside mud balls show that these mud balls are not represented by any genera of foraminiferan group. These mud balls are not armoured with any shells, granules or pebbles like of those mud balls as described by Bell (1940), Loney and Loney (1957) and Kugler and Saunders (1956). Occurrences of these mechanically grown up mud balls have less geological significance of the recent past but throws some light on the computation of rate of erosion of the beach environment that helps in the interpretation of morpho-dynamics of the coastal environment.

5.12 Trash in the Beach

Bakkhali at the bosom of the Bay of Bengal is almost lonely sea resort situated about 139 km from the nearest mega city Kolkata. Yet when a survey conducted on July 16, 2009, about 1,000 pieces of trash was noticed in 1.2 km stretch of this beach. Non-biodegradable trashes particularly the plastic end up in the sea coming from the lands. All these pose hazards to the marine life specifically when the animals drown or strangle from getting tangled in discarded or lost fishing gear or suffer and even die from swallowing plastic and other garbage. Plastic packets look like jelly fish are taken for feeding by the sharks and turtles.

5.13 Coastal Waters

The deleterious effect of water pollution has also been recorded as the pristine beach of Bakkhali. Chemical Oxygen Demand (COD) and Total dissolve solid (TDS) of the coastal water samples of the Bakkhali beach cross the permissible limit. Spatial variations of some physicochemical properties of the coastal waters of

Bay of Bengal at Bakkhali were measured. The mixing of river waters from Hugli flowing through Haldia industrial belt (basically Chemical detergents, batteries, petrochemicals and plastics) in coastal areas are the primary causes of nutrient enrichment, hypoxia, harmful algal blooms, toxic contamination, sedimentation and other problems that plague coastal waters of Bakkhali Beach.

5.14 Sedimentation at Ganga Sagar Beach and Chuksar Island

The Gangasagar beach, south of Sagar Island-an island stands along the coastal stretch of the Bay of Bengal and the Chuksar Island – an offshore island in the Bay of Bengal is 16 km away from the Gangasagar beach (Fig. 5.6). The bed

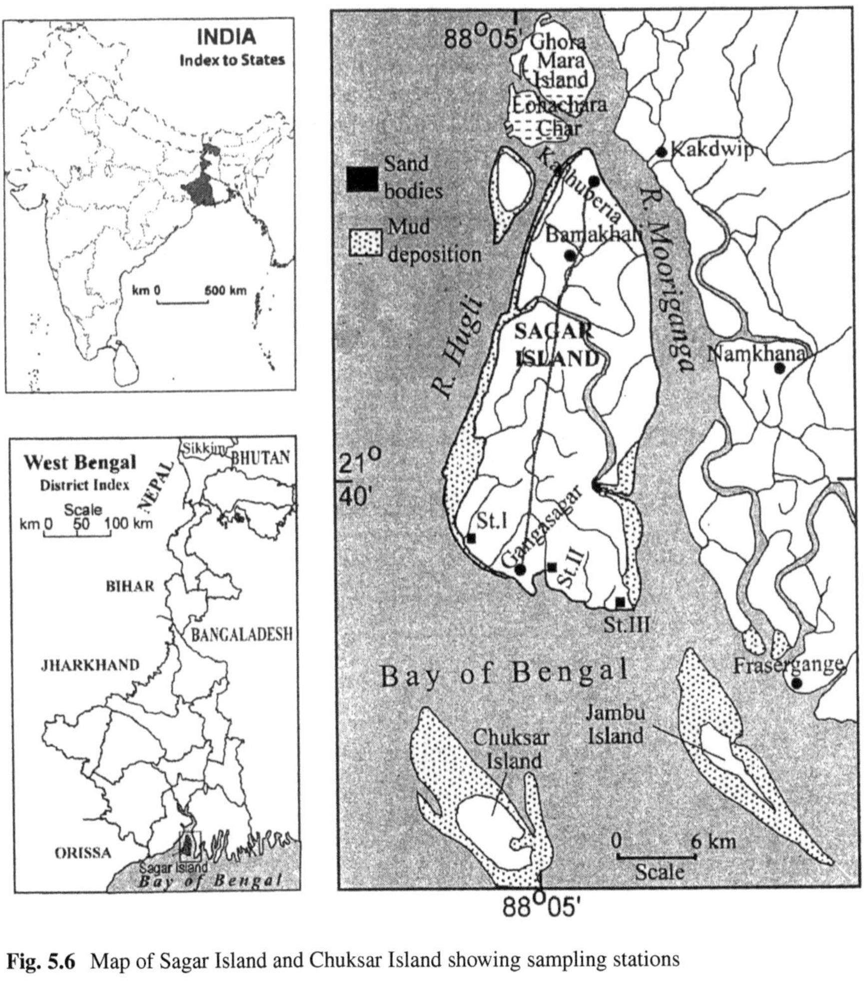

Fig. 5.6 Map of Sagar Island and Chuksar Island showing sampling stations

forms occurring in these generally sandy intertidal zones have been examined with special reference to their (i) association and distribution in space, (ii) relation to flow phenomena and (iii) identification of the different intertidal zones from low to high water level based on their association.

The fluctuation of tides in the form of rise and fall of water level causes migration of energy zones, which results characteristic surface sedimentary structures dependent on factors like beach slope, local morphology, depth of water, flow direction, flow velocity and textural attributes of sediments. Apropos of the variations of these elements on the beach surface, the surface sedimentary structures show wide variations in their morphological forms and dimensions within small locales of intertidal zones. Lateral variations of these structures as well as their variations at right angles to the coastline have been studied to interpret the hydrodynamic conditions of sedimentation.

The upper foreshore zone is characterized by the occurrence of megaripples marks superimposed with the small crested ripple marks (Fig. 5.7), sinuous trains of backwash ripple marks, swash marks and rhomboid marks. The middle foreshore zone with runnels and ridges show assemblages of lingoid ripple marks of large and small scale, interference ripple marks, particularly of polygonal and ladder back type, rill marks, rhomboid marks and water marks on lee and stoss slopes of large scale ripple marks. The lower foreshore zone, however, is characterized by the capped-off ripple marks, ripple- rill complex, double-crested ripple marks and current crescents.

These bed forms have been studied in the sandy beach, South Sagar Island and the Lower Long Sand – an offshore tidal island in the Bay of Bengal, 16 km South of Gangasagar beach (Fig. 5.6). These bed forms have been examined with special reference to (i) their distribution in the natural settings in the beaches, (ii) their relation to flow-phenomena and (iii) identification of the different beach zones from low to high water level based on their association.

Fig. 5.7 Megaripples marks superimposed with small crested ripples marks (*left*) and Rhomboid marks in the intertidal zone of Bakkhali (*right*)

5.15 Bedforms of Upper Foreshore Zone

5.15.1 Backwash Ripple Marks

These are parallel to shore gentle undulations very much unlike the aqueous ripples. The ripples are sinuous in plane and are typical of the upper foreshore zone, particularly in the regions of maximum advance of the wave swash. So they are invariably associated with swash marks and rhomboid marks. These ripples are concentrated within a width of 13–16 m. The wave length of these ripple marks ranges between 20 and 35 cm whereas the height is very low to be measured in mm. Concentration of dark heavy minerals, mostly biotite are noticed just behind the narrow-leading edges of light minerals composed chiefly of quartz and feldspar.

These ripple marks have been interpreted either as antidunes or as bedforms formed beneath undular hydraulic jumps produced when a backwash flow collides with incoming surf bore.

5.15.2 Rhomboid Ripple Marks

Associated with or superimposed on the backwash ripples in many occasions, are the rhomboid ripple marks (Fig. 5.7) variously known as rhomboid marks, rhomboid rill marks (Otvos 1964). These ripples appear with reticulate or tongue like crestal patterns. The width ranges between 4 and 6 cm. The length – width ratio (L/W) ranges between 1.56 and 2.9. Generally the length – width ratio maintains a direct relation with the beach slope. It is a statistical fact that with increase in beach slope, the L/W also increases. Similar observations on the rhomboid marks are obtained from Hoyt and Henry (1963).

The upper foreshore zone along with backwash ripples and rhomboid ripples (Table 5.2) are associated with current lineated plane beds and thus reflect association of bed forms of high power.

Table 5.2 Assemblages of bed forms typical for the zonation of beach of Sagar Island

Lower foreshore	Mid foreshore	Upper foreshore
Double crested ripple marks	Mega ripples (lunate)	Backwash ripple marks
Ripple-rill complex, capped off	Sandy ribbon	
Ripple marks, ripples with slope	Lingoid ripple marks	Swash marks
(Trough line)	Ripple toroid	Rhomboid ripple marks
	Interference ripple marks	
	(Ladder back, polygonal, Tadpole nests)	
	Watermarks	
	Rill marks, current crescent	

5.16 Bedforms of Mid Foreshore Zone

The mid foreshore zone of the beaches are often marked by runnel and ridge topography which comprises a series of coast parallel asymmetrical ridges separated by shallow troughs or runnels 50–100 m wide. It has been suggested by Reading (1984) that such a rhythmic topography is favoured by moderate wave energy conditions acting on fine grained meso-tidal to macro-tidal beaches with abundant sediment supply. These ridges are supposed to be originated by swash backwash processes during storms. Beach cusps are parallel to the coast constitute the other types of rhythmic topography in the mid foreshore zone. These runnels on the beaches are the areas of sculpture by mega ripple marks lingoid ripple marks and interference ripple marks.

5.16.1 Mega Ripple Marks

The mega ripple marks in the present field are always larger than 50 cm in wavelength and ranges up to 80 cm or slightly more. The megaripples always maintain a separation in dimension from the small ripples. They usually developed under higher energy conditions than that of ripple marks and always located on the beds of the shallow tidal creeks and rubbles where the water height is generally higher than other beach areas. The megaripples migrate along the direction of flow of water inside the runnels and creeks and their orientations are always controlled by the flow of water through the creek and runnels.

With a decrease in flow velocity, the mega ripples cease migration and the effect of wave-swash and backwash generates small-scale ripple marks superimposed on them. The plots of wavelength and height of mega ripples and that of superimposed small ripples show no overlapping in a single occasion. The effect of continuous fall of water level during recession of tides, gives rise to water marks that are seen on the lee and stoss slopes of mega ripples and sometimes on sloping face of the ripples. The lunate mega ripples are more common compared to the lingoid mega ripples. The lunate type have discontinuous crests which are in general crescentic in shape, sometimes they are D shaped in plan and often form spoon shaped scours (Harms and Fahenstock 1965).

5.16.2 Lingoid Ripple Marks

These are formed on the beach runnels, on the shallow tidal creek or on the floors of gulley-like depressions emerging at right angles from runnels. Lingoid ripples form by unidirectional movement of water along elongated shallow depressions or runnels. With decreasing beach slopes the lingoid ripples grade to small undulatory ripples, rhomboid ripples and into many transitional forms (Fig. 5.8).

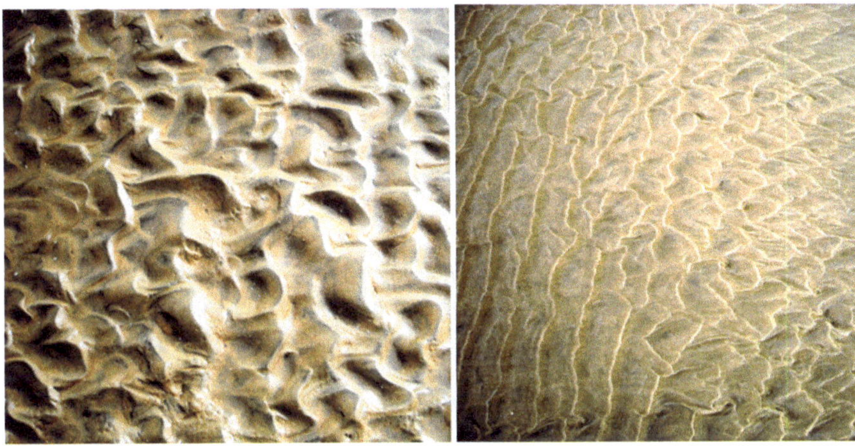

Fig. 5.8 Lingoid ripple marks (*left*) and straight crested small ripple marks tends to interference and polygonal ripples (*right*)

5.16.3 Interference Ripple Marks

Interference ripple marks (Fig. 5.8) have been named variously by different workers depending on their morphological varieties as – polygonal ripples, Ladder-back ripples, cross ripples and tadpole nests.

5.16.3.1 Polygonal Ripple

In the present situation, interference ripples of polygonal type have of pentagonal crests (Fig. 5.8) and have been observed particularly where a complex interference of the waves and currents takes place in certain portions of the tidal channels crossing through the Gangasagar beach. Risto Aario (1971) have described polygonal ripple pattern from the Otsker delta of Finland.

5.16.3.2 Ladder – Back Ripples

Due to flooding and ebbing tides, water level fluctuates on the foreshore zone and this causes the tidal currents responsible to initiate the tidal channels on the foreshore beach zones. At the first phase of falling water level, the water flows parallel to the shore over both crest and troughs of ripples. This creates a set of current ripples marks over the crests and troughs of the larger wave ripples. At successive lowering stages of the water level the water gets confined within the troughs of the earlier set of ripple marks its movement through the troughs creates a second set of ripple marks confined within the troughs of the earlier ripples. The superimposed sets of ripple marks together are termed ‘ladder-back’ ripple marks and these ripples abound in the tidal channels of both the sandy beaches and in the mud flats of the Harinbari area of Sagar Island. The plots of wavelength and height of the two sets ripple do not show any overlap.

5.16.3.3 Cross Ripples/Tadpole Nests

Hitchcock (1958) however describes some interference ripples under the name of ‘tadpole nests’ after the supposed origin but Bucher (1919) calls these structures as cross ripple. It has been suggested that these structures originate in shallow water through the intersecting pattern of simultaneous or successive waves or waves subsequent upon currents. Thus, they are supposed to be closely related to ordinary waves (oscillation) ripple marks.

5.16.3.4 Arcuate Sands Resembling Lunate Megaripples

Strong winds sometimes pucker the water of puddles and move it in the form of arcuate streaks. These arcuate streaks are discontinuous in plane and often resemble incomplete lunate mega ripples. Similar structures have also been described by Reineck and Singh (1980).

5.17 Bedforms of Lower Foreshore Zone

5.17.1 Double Crested Ripples

These ripples are more or less wavy to straight in plane and are characterized by double crests. The double crested characters of these ripples are seen exclusively at the junction where the vigour of wave swash and backwash counter balances each other.

The wave swash at the time of rushing up over the mid foreshore zone of the beach creates ripple crests with lee slope pointing against sea. The backwash vigour of flow over the lower foreshore zone is capable to produce ripple lee-slope directed towards the sea. The effects of swash and backwash are so balanced in the zone transitional between the lower and mid foreshore that the ripples attain a more or less symmetrical profile and their effects leave two close-by parallel to sub parallel crestal marking to give rise to a double crested pattern of the ripple marks.

It is important to note, however, that these double crested ripples are unlike the double crested ripples of Reineck and Singh (1980) where they suggest their origin from falling water level of high water stage.

5.17.2 Straight Crested Small Ripples Marks

These ripple marks are characteristic of the lower energy zones of beaches (Fig. 5.7) and are mostly found in areas where the water height is very low and flow velocity is very gentle. These straight crested ripples often grade to symmetrical wave ripples. These ripples are often associated with small rills.

5.17.3 Capped – Off Ripples

These ripples are typical of lower foreshore zone. Capping-off or flattening off the crest of the Ripples takes place during sub-aerial emergence of beaches. According to Reineck and Singh (1980), this flattening is caused by capillary waves that are produced by strong winds on water surfaces. Here the ripple crests become rounded and lowered and the sand flows the crest into the troughs of ripples. This causes, the ripple troughs to become with very sharply pointed grooves.

5.17.4 Ripple – Rill Complex

After the recession of the tide from the lower foreshore zone, the water-saturated sediment of that particular beach is responsible to generate structures like ripple-rill complex (Fig. 5.9). The water that is present in the trough of the ripple marks as well as water that is expelled from the sediments during the emergence of the lower foreshore zone is capable to produce similar rills interlacing with ripple marks.

5.18 Texture of Beach Sediments

Textural analyses of beach samples (six in numbers) taken from different kinds of ripple marks are presented in Table 5.3. The statistical size parameters calculated often Folk and Ward (1957) show that the sediments are generally well-sorted fine sandy in nature. All these samples are slightly positively skewed and show mostly leptokurtic ($K_G > 1.00$) nature (Table 5.3). The cumulative curves (Fig. 5.10) are mostly non-linear with the saltation mode of transport as the chief population for the distributions. Inflections at 2.5–3.0 phi are a statistical fact for all the samples. All the curves show close resemblance in pattern with each other (Fig. 5.10).

Fig. 5.9 Ripple-Rill complex at the lower intertidal zone

Table 5.3 Statistical grain size parameters of the sediment samples taken from ripple marks of various types

Sample no.	Graphic mean size (M_Z)	Inclusive graphic standard deviation	Inclusive graphic Skew ness (SK_1)	Inclusive graphic kurtosis (K_G)
S_1	2.783	0.304	0.292	1.850
S_2	3.168	0.470	0.150	0.746
S_3	2.967	0.452	0.478	2.077
S_4	3.00	0.480	0.362	1.096
S_5	1.90	0.351	0.291	1.447
S_6	1.734	0.255	0.027	1.295

5.19 Beach Character and General Flow Phenomena

Swash activity on the intertidal zone of the beaches allows large quantities of water to move through the beach sand and a water depth of 2 m acts upon beaches at their lower foreshore zone. Further the fluctuation of tides in the form of rise and fall of

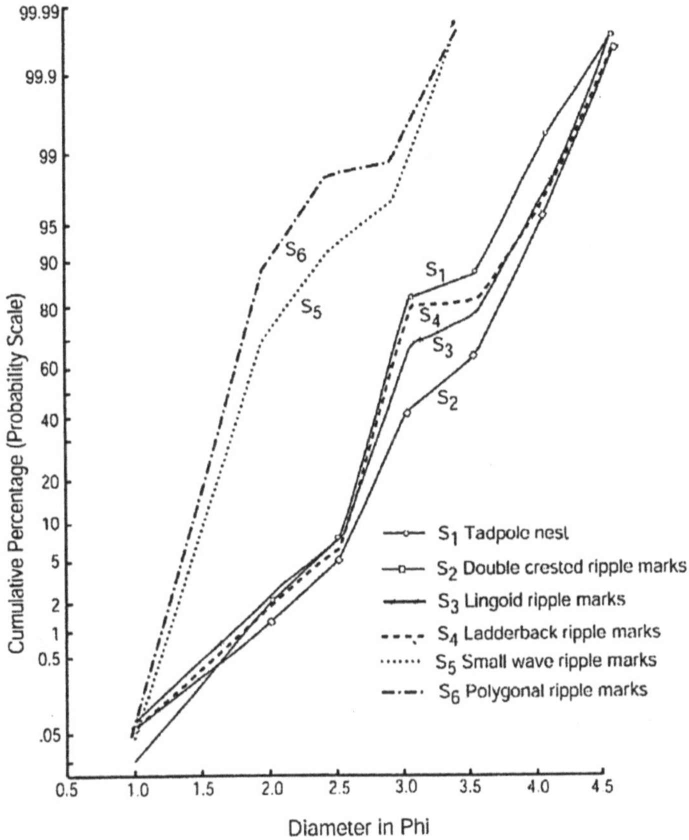

Fig. 5.10 Cumulative curves for the various ripple marked samples

water level causes migration of energy zones which result characteristic sedimentary structures dependent on the important factors like beach slope, local morphology, the water depth, flow velocity, and grain size and sorting. As all these parameters show variations within small stretches of areas the sedimentary structures also show wide variations in these morphological characters and in the scales of their formation within small areas of the beaches. Lateral variations of these structures and their up and down translation along beach profile with change of energy zone have been studied to interpret the hydrodynamic conditions of sedimentation (Das 2004). The ripple marks occurring along beach profile from the lower to the upper foreshore zone generally show an increase in the strength of flow belonging to the lower flow regime (Fig. 5.11).

The beach south of Sagar Island facing Bay of Bengal has a general slope ranging between 2° and 4° with the maximum of 6°. The beach face however shows a greater angle to the order of 14°–16° on slopes of the ridges or long shore bars that run parallel to shore.

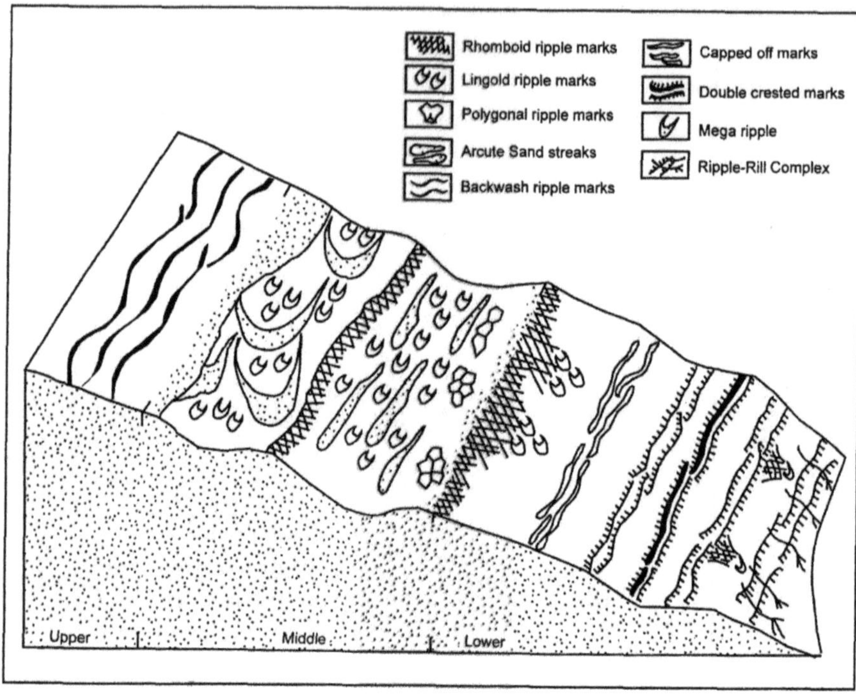

Fig. 5.11 Diagrammatic sketch of the Gangasagar beach profile showing occurrence of various bedforms in different parts of foreshore zone

The beach profile of Lower Long Sand on all sides of the island is much more convex from the backshore to the lower foreshore region compared to that of the Gangasagar beach. The beach slope here ranges between 2° and 12°.

Along the beach profiles of these two islands, a more or less systematic variation of the bed forms can be noticed (Table 5.2). The upper foreshore zone is conspicuously characterized by the occurrence of sinuous trains of backwash ripples and rhomboid marks. Profusely occurring ripple structures of mega and small scales of lingoid and interference types characterize the middle foreshore zone with runnels and ridges. Double crested ripples, straight crested to sinuous trains of ripples, capped-off ripples, ripple-rill complex, characterize the lower foreshore zone (Fig. 5.11 and Table 5.2).

5.20 Summary

Sea level rise due to climatic change may have a considerable impact on the coastal Sunderbans because it stands at the lower deltaic plain of the Ganga – Brahmaputra delta and is a low-lying coast. Reduction of beach width and alteration of coastal geomorphic situations are the consequences of the slight rise of the sea level (Pethick 2000; Klein et al. 2001) in such a low lying coast like Sunderbans which is geologically soft and susceptible to erosion. Erosional features help in understanding the morphodynamic processes of this sea beach. This erosion leads to the changing configuration of this coastal zone under the influence of variable coastal agencies. The morphodynamic changes along this coast are controlled by catastrophes like cyclonic storms, aeolian, fluvial, tidal and fluvio-tidal events. Erosional sedimentary structures like mud mounds, erosional furrows and ridges are transverse to the strand in the intertidal shore along Bakkhali stretch of West Bengal. Out of these erosional structures, obstacle scours are important which are abundantly occurred in the upper and middle intertidal zone. Obstacle scours, generally eroded wood trunks and mud balls, help in resisting erosion because of obstructions during recession of water over the foreshore regions. Sometimes troughs are formed round to the obstacles due to vigorous tidal flows during flood tides. Gradual enhancement of landward displacements of the shoreline of Bakkhali is accelerated as inferred from field evidences. The Bakkhali intertidal area has faced two different types of disturbances-(i) the sand deposits are removed and transferred seawards by waves and tides and (ii) scouring in the beach erodes mud and that muddy sediments are carried away to the sub-tidal region by suspension transport (Bhattacharya 2000). This sediment load helps in the formation of tidal shoal in the sub tidal region. In a beach face there is a general trend of successive variations of ripple type along with other associated surface structures. The lower limit of the foreshore zone characterized by oscillatory flow is seen to be replaced by symmetrical flow more and more from mid foreshore to upper foreshore region. This reveals an increasing flow power as registered and reflected from the changing in the pattern of bedforms (Table 5.2). Moreover, these bed forms indicate the flow regime change (Harms and Fahenstock 1965) in the beach face which starts from lower part of the lower flow regime in the lower foreshore zone through the upper part of the lower flow regime in the mid foreshore zone to the transitional to upper flow regime in the upper foreshore zone.

References

Allison MA (1998) Geological framework and environmental status of the Ganges- Brahmaputra delta. J Coast Res 14(3):826–836

Bell HS (1940) Armoured mud balls-their origin, properties and role in sedimentation. J Geol 48:1–31

Bhattacharya A (1988) Intertidal depositional characters of the Hooghly tidal islands, West Bengal. Indian J Geol 60(3):153–164

Bhattacharya A (2000) Some geomorphic observations indicating shoreline variation in the coastal tract of West Bengal (India) – a case study around Bakkhali in Sunderbans. In: Proceedings of the international quaternary seminar on INQUA shoreline. Indian Ocean Sub-Commission, New Academic Publishers, Delhi, pp 32–37

Bucher WH (1919) On ripples and related surface forms and their paleo-geographic interpretations. Am J Sci 47:149–210

Carle L, Hill PR (2009) Subaqueous dunes of the upper slope of the Fraser River delta (British Columbia, Canada). J Coast Res 25(2):448–458
Carranza ML, Carboni M, Feola S, Acosta ATR (2010) Landscape-scale patterns of alien plant species on coastal dunes: the case of iceplant in central Italy. Appl Veg Sci 13:135–145
Coleman JM (1969) Brahmaputra River: channel process on sedimentation. Sediment Petrol 3((2/3) Spl. Issue):129–139
Das GK (2004) Morpho-dynamics of deltaic Sunderbans Rivers. In: Singh S, Sharma HS, De SK (eds) Geomorphology and environment. Acb Publications, Kolkata, pp 303–308
Das GK (2009a) Beach processes of Tidal Islands of Hugli Estuary, West Bengal. Geogr Rev India 71(3):240–248
Das GK (2009b) Grain size analysis of some beach sands from the Indian Coasts. Geogr Rev India 71(1):10–18
Das GK (2010) Environmental status of Bakkhali Beach, West Bengal. Geogr Rev India 72(2): 132–139
De Boer PL, Oost AP, Visser MJ (1989) The diurnal inequality of the tide as parameter for recognizing tidal influences. J Sediment Petrol 59:912–921
Folk RL, Ward W (1957) Brazos River bar – a study of the significance of grain size parameters. J Sediment Petrol 27:3–26
Friend PL, Caviola P, Cappuchi S, Santosh R (2003) Bio-dependent bed parameters as a proxy tool sediment stability in mixed habitat intertidal areas. Cont Shelf Res 23:1899–1917
Harms JC, Fahenstock RK (1965) Stratification, Bed forms and flow phenomena with an example of Rio grade in Primary Sedimentary Structures and their hydrodynamic interpretation, S.E.M.M. Sp. pub. 12, pp 84–115
Hitchcock E (1958) Ichnology of New England, a report on the sandstones of the Connecticut valley. William White, Salte Printer, Boston
Hoyt JH, Henry VJ (1963) Rhomboid ripples marks, indicator of current direction and environment. J Sediment Petrol 33:604–607
Klein RJT, Nicholls RJ, Ragoonaden S, Capobianco M, Aston J, Buckley EN (2001) Technological options for adaptation to climate change in coastal zones. J Coast Res 17:531–543
Kuehl SL, Levy BM, Moore WS, Allison MA (1997) Subaquaous delta of the Ganges- Brahmaputra river system. Mar Geol 144:81–96
Kuglar H, Saundars P (1956) Occurrence of armoured mud balls in Trinidad, West Indies. J Geol 67:563–565
Loney LB, Loney RL (1957) Glacial outwash plains. J Geol 65:310–316
McKee ED (1957a) Primary structures in some recent sediment. Am Assoc Petrol Geol Bull 41:1704–1747
McKee ED (1957b) Flume experiments on the production of stratification and cross stratification. J Sediment Petrol 27:129–134
Milliman JD, Meade RH (1983) World wide delivery of river sediments to the oceans. J Geol 91:1–22
Mitra PK, Samaddar AK (2005) Erosion in the Frasergunj-Bakkhali sector, South 24 Parganas, West Bengal. News Lett GSI Mar Wing XXI (1):30–32
Morgan JP, Mclwtire WG (1959) Quaternary geology of the Bengal Basin, East Pakistan and India. Geol Soc Am Bull 70:319–342
Otvos EG (1964) Observations on Rhomboid beach marks. J Sediment Petrol 34:683–687
Pethick J (2000) An introduction to coastal geomorphology. Arnold Publ., London
Reading HG (1984) Sedimentary environments and facies, 2nd edn. Blackwell Science Publications, Oxford
Reineck HE, Singh IB (1980) Depositional sedimentary environments. Springer, New York
Risto Aario (1971) Association of Bed forms and paleo-current pattern in an Esner delta, Haapajarvi, Finland. Annales Academiae Scientiarum Fennicae. Series A.III, Geologica – Geophysica

Rose PR, Lidz B (1977) Diagnostic foraminifers' assemblages of shallow-water modern environments – South Florida and the Bahamas. In: Lidz B (ed) Sedimenta VI, Comparative Sedimentology Laboratory, Division of Marine Geology and Geophysics, University of Miami, Miami

Roy Barman A (1983) Geology and hydrocarbon prospects of West Bengal. Petrol Asia J 1:51–56

Roy Barman A (1992) Geological history and hydrocarbon exploration in Bengal Basin. Indian J Geol 64(3):235–258

Van der Maarel E (2003) Some remarks on the functions of European coastal ecosystems. Phytocoenologia 33:187–202

Vila' M, Weber E, D'Antonio CM (2000) Conservation implications of invasion by plant hybridization. Biol Invasions 2:207–217

Vila' M, Tessir M, Suhes CM, Brundu G, Carta L, Galanidis A, Lambdon P, Manca M, Me'dail F, Moragues E, Traveset A, Troumbis AY, Hulme PE (2006) Local and regional assessments of the impacts of plant invaders on vegetation structure and soil properties of Mediterranean islands. J Biogeogr 33:853–861

Chapter 6
Granulometry of Mangrove Sediments

Abstract The grain size analyses of estuarine sediments give an idea of the depositional environment in the estuaries. Values of graphic mean size, median, standard deviation, kurtosis and skewness are determined from the cumulative curves drawn by plotting size values. The statistical size parameters of the Sunderbans estuarine sediments indicate that the estuarine banks, flanks of the mid channel bars and point bars where depositional energy is low, are characterised by muddy sediments.

Keywords Sunderbans estuary • Grain size • Mean • Median • Standard deviation • Kurtosis • Skewness • Cumulative curves

Sediment is made up of loose particles of sand, silt and clay. Particle size refers to the diameter of individual grains of sediment. It is a fundamental descriptive measure of sediments from any environment. Grain size analysis of estuarine sediments is required to study the trends in surface processes related to dynamic conditions of transportation and deposition.

Relative proportions of sand, silt and clay determine the soil's texture. Grain size of estuarine sediments is the characteristic feature of the size frequency distribution of sediments as well as it is a fundamental descriptive measure of sediments in the estuarine environments. It helps in the interpretation of depositional environment of clastic sediments of a deposit by parameters of their texture. The general physical appearance of an object is considered as texture. Estuarine environment is characterised by the texture of its sediments which has a relationship between texture of sediments and process of deposition. As a whole, sediment grain size is an important aspect which is related to the transport and deposition of sediments.

The average size of the grains of materials in a sample also known as fragments or particle size and the general physical appearance of an object is grain Size where sand particle size varies from 0.0625 to 2 mm. Sand size class ranges from −1.0 to 4.0 (phi). Silt is a particle size of 0.008–0.0625 mm. It is intermediate in size between sand and clay. Silt size class ranges from 4.0 to 8.0 (phi). Clay is a particle size between silt and colloid of various hydrous aluminum silicate minerals that are plastic, expansive, and have ion –exchange capacities. Clay size class ranges from 8 (phi) and onwards.

G.K. Das, *Estuarine Morphodynamics of the Sunderbans*,
Coastal Research Library 11, DOI 10.1007/978-3-319-11343-2_6

6.1 Measures of the Grain Size Distribution

The nature of grain size distribution in sediments of estuarine or any environment can be described on the basis of five specific parameters. The graphic mean size is an arithmetic average of a series of diameter values. The median diameter is the 50th percentile diameter of a cumulative frequency curve drawn on arithmetic probability paper. Standard deviation is expressed as measures of dispersion (sorting) of sediments and it is the square root of the arithmetic average of the squares of all the deviations from the mean size value of a series of observations. Skewness measures the asymmetry of the grain size distribution. Grain size distribution is skewed when the mean deviates from the median. Skewness of the sediments for symmetrical grain size distribution is zero. Skewness becomes negative when the grain size is skewed towards smaller phi value, and it is positive when skewed towards higher phi value. Kurtosis is the condition of peakedness or flatness of the graphic representation of a statistical distribution.

6.2 Statistical Analysis of Grain Size

Texture of muddy sediments of estuaries is examined by mechanical analysis of sediments following sieving-cum-pipetting method and sand and gravel fractions were determined by sieve analysis (Krumbein and Pettijohn 1938) using sieves of different mesh sizes marked as ASTM (American Society for Testing and Materials). The statistical size parameters are calculated using formula of Folk and Ward (1957) from the cumulative curves drawn on arithmetic probability papers. Later, rapid sediment analyzers (RSA) propounded by Zeigler and others (1960) and Schlee (1966) and electro-resistance multichannel particle-size analyzers (EMPSA) were introduced for automated analysis and calculation of statistical parameters of sediments. Contemporaneous with them Kane and Hubert (1962) and Schlee and Webster (1967) developed Formula Translation programmes (FORTRAN) for textural analysis of sediment particles parameters. Gradually, Algorithmic Language (ALGOL) by Jones and Simpkin (1970), Beginners All-Purpose Symbolic Instruction Code (BASIC) by Sawyer (1977) and hand-held calculators by Benson (1981) were programmed for statistical grain size computations. About the same times, hardware and software packages introduced by many workers (Muerdter et al. 1981; Poppe et al. 1985) and organization (Coulter Electronics 1989) for electro-resistance multichannel particle-size analysis were introduced. Introduction of computer driven integrated particle size analysis instruments fitted with settling tube (Rigler et al. 1981) automated and modernized the sediment grain size analysis method. Settling tube also called rapid sediment analyzer design based on using the pressure differential between two columns of water that have a common head provides for efficient analysis of sand-sized material by setting the grains where results are relayed to a personal computer associated with data acquisition software drivers (Syvitski et al. 1991). A computer programme called

GRADISTAT (Blott and Pye 2001) has been written for the rapid statistical analysis of size data from any standard measuring techniques. The programme runs with Microsoft Excel package. It is very much useful and produces a range of graphical representation including frequency curves and plots.

6.3 Sampling of Sediments

It needs utmost care during sampling of estuarine sediments because the grain size analyses are sensitive to the manner in which the original samples are collected, handled and preserved. Introduction of any foreign particle into the sample through improper care, cleaning of equipment or processing, can alter the texture. Estuarine landforms like point bars, river mouth bars, tidal shoals, major tidal inlets, upstream and downstream of the rivers are ideal sites for collection of samples. Data regarding tide, current, waves, depth, turbidity etc. are also collected during sampling of estuarine sediment for size analysis. Instrumental tripod ALICE fitted with various sensors is used to collect data regarding the above physical parameters. Numerical models are used for data interpretation.

6.4 Cumulative Curves

Cumulative curves plotted on arithmetic probability paper represent grain size distributions of different sub-populations which has a log normal distribution depicting different modes of transportation of sediments (Visher 1969). Sediment grain size is determined from grain size curves drawn on log-probability plots. It helps in the interpretation of separate populations of estuarine sediments. Three different methods of plotting are considered for grain size distribution such as grain size with frequency percent, cumulative frequency percent and the log-probability cumulative frequency percent. The log-probability cumulative frequency curves is the most accepted methods by the sedimentologists for depositional environment of the estuaries. In each log-probability curves there are at least four control points i.e. four separate log-normal populations, where each population is truncated and the former one joined with the later one to make a single grain size distribution. Each log-normal population is composed of different means and standard deviation and those populations are identical with log-probability plot.

6.5 Expression of Sediment Grain Size

Estuarine waters transport a wide range of sediments varying in size from 2 μm (0.002 mm) to more than 4 mm, but finer sizes dominate most estuaries. A few estuaries transport sand (>62 μm), gravel and larger sediments.

Sediment grain size is expressed as phi (ϕ) by Krumbein (1938) as logarithmic diameter has more significance in a discussion of the statistical relations of sediments. Sediment grain size in phi (ϕ) is expressed as the negative logarithm to the base 2 of the sediment particle diameter in mm. Thus sediment grain size is expressed as following:

$$\phi = -\log_2 \varepsilon$$

Where ϕ stands for sediment grain size and ε equals to negative numerical value of diameter. ε equals to 2, 1, ½, ¼ etc., whereas ϕ equals to -1, 0, $+1$, $+2$ etc. Thus ϕ increases with decreasing diameter.

6.6 Grain Size Characteristics of Different Estuarine Landforms

There is an interrelationship between grain size characteristics and depositional pattern in a tide dominated estuarine environment. Interpretation of the grain size frequency curve is based upon the pattern of curves and splitting of each curve into segments separated by the marked breaks and inflection. Sediments from different geomorphological areas like point bar, mid channel bar, swash bar, river bank and areas of other morpho-ecological interests may be considered for grain size analysis.

6.7 Grain Size Populations

Some cumulative curves show rhythmicity in the nature of deposition, which perhaps indicates the depositional pulses for the ebb and flood flows. The inflection points in the curves between successive ebb – flood cycles are marked at 1.5–2.0 phi, 2.5–3.0 phi, 3.5–4.0 phi respectively. Such rhythmicity in the nature of depositional behaviour of the tide influenced beach sands is supposed to be process responsive. The cumulative curves for the sediment samples from Bakkhali show a non-linear pattern. These samples show two major inflections one at 2.25 phi and the other at 3.0–3.5 phi. These inflections divide the curves into three subpopulations as rolling, saltation and suspension respectively (Fig. 6.1). The saltation population constitutes about 75 % of the total material, the rest being deposited by either rolling or suspension. The central saltation population constitutes about 70 %. These sediments do reflect their deposition from tractive movements of water in a hydrodynamic flow condition. Fuller (1961) suggested on the log probability plots that the break between saltation and rolling populations in many instances occurred near 2 phi. Whereas, Spencer (1963) suggested that: (i) all clastic sediments are mixtures of three or less log – normally distributed populations; and (ii) sorting is a measure of the mixing of these populations (Visher 1969). Visher (1969)

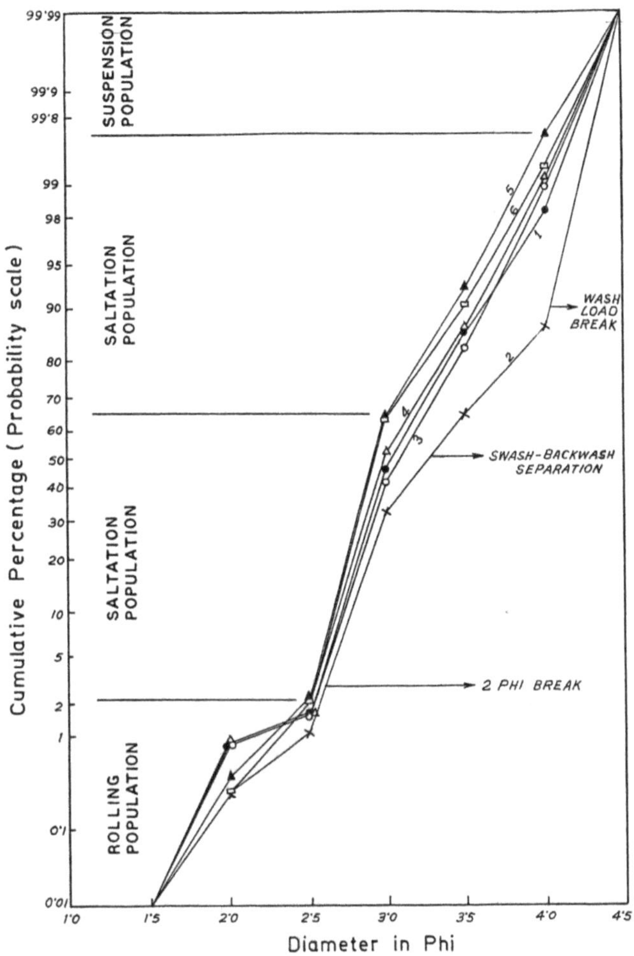

Fig. 6.1 Cumulative curves of the sediment samples of Bakkhali showing different grain size populations

suggested that: (i) suspension, saltation and surface creep are three modes of sediment transports and (ii) there is no vertical change in grain size occurs where true suspension caused by turbulence.

6.8 Texture of Dune Sands

The grain size of dune sands from Fraserganj area generally show log normal distribution having very fine sand and well sorted sediments. The graphic mean size ranges between 2.5 and 3.5 phi (Fig. 6.2). The graphic standard deviation indicates

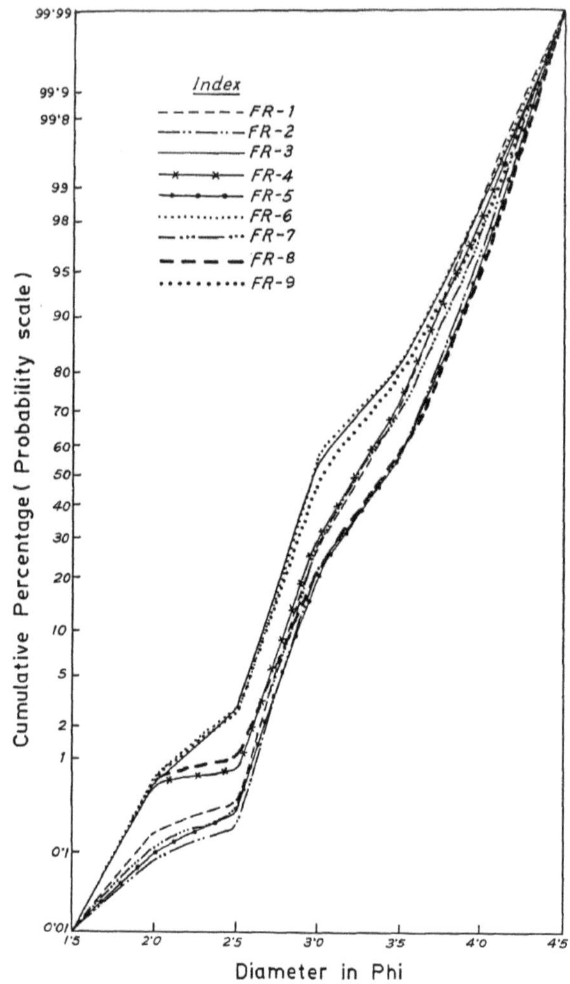

Fig. 6.2 Cumulative curves for the dune sediment samples of Fraserganj

well sorted finer sand particles. The sorting value reflects the reworked intertidal zone sediments through wind transportation cycles. Dune sediments show a slightly positive skewness because of admixture of very small quantity of finer suspended particles to the log normal saltation population. Kurtosis of sediments is very close to 1.0 indicating log normal distribution with mesokurtic characteristics (Table 6.1).

6.9 Texture of Beach Sediments

Textural analyses of beach samples (six in numbers) taken from different kinds of ripple marks are presented in Table 6.2. The statistical size parameters calculated after Folk and Ward (1957) show that the sediments are generally well-sorted

Table 6.1 Statistical grain size parameters of the dune sands of Fraserganj

Sample no	Graphic mean size (Mz)	Inclusive graphic standard deviation (σ_1)	Inclusive graphic skewness (SK_1)	Inclusive graphic kurtosis (K_G)	Sand %	Silt %	Clay %
F_1	3.22	0.43	0.01	0.91	100	–	–
F_2	3.35	0.44	0.16	0.81	100	–	–
F_3	3.03	0.41	0.27	0.86	100	–	–
F_4	3.23	0.43	0.01	0.88	100	–	–
F_5	3.36	0.44	−0.18	0.81	100	–	–
F_6	3.02	0.41	0.31	0.86	100	–	–
F_7	3.22	0.43	0.03	0.87	100	–	–
F_8	3.32	0.44	−0.12	0.81	100	–	–
F_9	3.05	0.42	0.31	0.83	100	–	–

Table 6.2 Statistical grain size parameters of the beach sands of Gangasagar

Sample no	Graphic mean size (Mz)	Inclusive graphic standard deviation (σ_1)	Inclusive graphic skewness (SK_1)	Inclusive graphic kurtosis (K_G)	Sand %	Silt %	Clay %
G_1	2.78	0.31	0.29	1.86	100	–	–
G_2	3.16	0.47	0.15	0.75	100	–	–
G_3	2.97	0.45	0.47	2.08	100	–	–
G_4	3.00	0.48	0.36	1.09	100	–	–
G_5	1.90	0.35	0.28	1.45	100	–	–
G_6	1.73	0.25	0.03	1.29	100	–	–

fine sandy in nature. All these samples are slightly positively skewed and show mostly leptokurtic ($K_G > 1.00$) nature. The cumulative curves are mostly non-linear with the saltation mode of transport as the chief population for the distributions. Inflections at 2.5–3.0 phi are statistical fact for all the samples. All the curves show close resemblance in pattern with each other (Figs. 6.3 and 6.4).

6.10 Texture of Sand Flats

The sediments of sand flats consist of well sorted 95 % fine to very fine sands in comparison to silt flats sediments having 10 % sand, 95 % silt and 5 % clay. Cumulative curves for the sand flats sediments are very similar in pattern (Fig. 6.5) and differ much from the silt flat sediments.

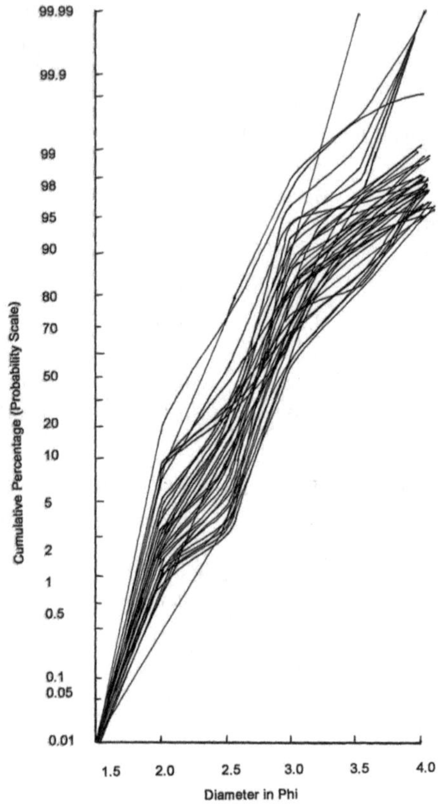

Fig. 6.3 Cumulative curves of the collected sediment samples

6.11 Texture of Tidal Shoal Sediments

Tidal shoals, in general, shows greater accumulation of mud in the upstream portion and sand in the downstream stretches. This characteristic depositional behaviour of sediments goes in favour of the tidal accumulation rather than its formation out of unidirectional flow from upper to the lower stretches. The downstream accumulation of sand suggests influence of flood flow, whereas, the upstream mud is the result of a stand still during the high and low tides.

The sediments are generally very well sorted to medium sorted fine silty in nature (Table 6.3). Samples are slightly negative to slightly positively skewed (SK_1) and are platykartic in nature (Fig. 6.5). Inflections generally at two to three truncation points are a statistical fact for all the samples. All the cumulative curves are mostly nonlinear showing close resemblance in pattern with each other.

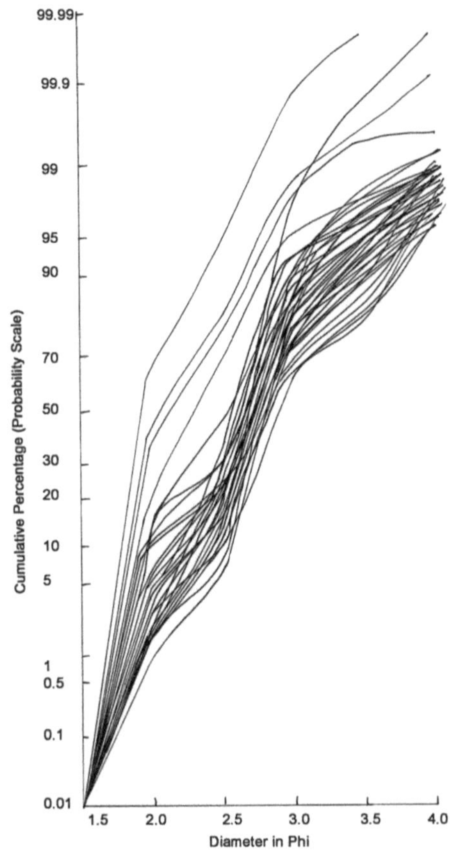

Fig. 6.4 Cumulative curves of the sandy sediment samples

6.12 Texture of Point Bar Sediments

Texturally, the sediments are generally composed of 95 % silt with sub equal proportions of fine sand and clay forming the rest 5 % of the remainder (Table 6.3). Sediments are mostly well sorted to moderately sorted with graphic mean size belonging mostly to the silt fraction. Most of the sediments show a slight tendency of negative skewness. The cumulative curves mostly show a higher peakedness with K_G values often close to 1.0 (Figs. 6.5 and 6.6).

Sandy sediments of the point bars show a completely different pattern from that of sandy silty sediments. The cumulative curves for these sediments show a non-linear pattern and are nicely comparable to that of the sandy point bar sediments. These sediments show two major inflections at 2.25 phi and 3.25 phi respectively. These inflections divide the curves into three subpopulations as rolling, siltation and suspension respectively. The saltation population constitutes about 75 % of the materials, the rest being deposited by either rolling or suspension. Silty sediments having range from 4 to 9 phi show exactly the same pattern where the three

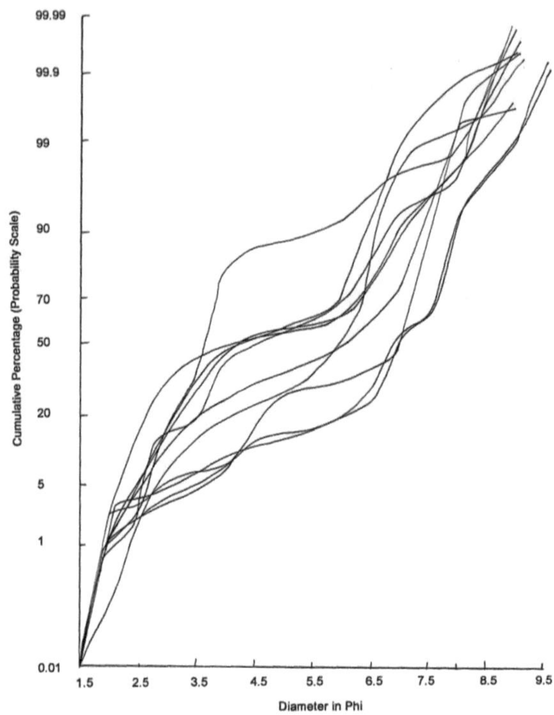

Fig. 6.5 Cumulative curves for the sediment samples

subpopulations can be well recognised. In these curves, the inflections take place at 6 phi to 7 phi respectively. The central saltation population constitutes about 75 %. These sediments do reflect their deposition from tractive movements of water in the estuarine flow condition.

6.13 Texture of Mudflat Sediments

The graphic mean size of surficial and depth samples lies within the silt fraction with moderately well to poor sorting. Sediments of mudflat samples show a positive skewness and reflect infiltration of suspended clay from tidal stand still through the pore spaces of the dominating silt and subordinate sand populations. Sediments of the mudflats traversing creek bottom, however exhibit negative skewness. This is because of the mixing of a greater proportion of sand fraction with the dominant silt population. The removal of clay with flowing creek water leaves the creek bottom with more sands compared to other places of the mudflat and this leads to a negative skewness of the distribution patterns.

Cumulative curves drawn from the mudflat sediments reveal close similarity in pattern (Fig. 6.6). The same is true for the creek bottom sediments when considered separately. The prominent breaks in the cumulative curves reflect changes in the mode of transport of suspended particles.

Table 6.3 Statistical grain size parameters of the river flood plain and mud flat samples

Sample no	Graphic mean sizes (Mz)	Inclusive graphic standard deviation (σ_1)	Inclusive graphic skewness (SK_1)	Inclusive graphic kurtosis (K_G)	Sand %	Silt %	Clay %
P_1	6.30	0.52	−0.26	2.51	4.2	87.9	7.9
P_2	5.58	0.76	−0.72	0.49	20.2	79.5	0.3
P_3	6.40	0.99	−0.81	1.58	7.4	92.0	0.6
P_4	6.32	0.56	−0.56	2.34	0.8	94.4	4.8
J_1	5.95	0.73	−0.39	0.96	11.7	87.9	0.4
J_2	5.90	0.84	−0.30	0.66	22.6	72.9	4.5
J_3	5.96	0.96	−0.15	0.93	13.1	85.4	1.5
J_4	5.77	0.63	−0.43	0.74	12.9	86.9	0.2
J_5	6.30	1.08	−0.30	0.69	18.5	77.7	3.8
J_6	5.97	0.94	−0.48	1.23	3.6	87.9	8.5
J_7	5.92	0.87	−0.62	0.80	4.5	85.4	10.1
J_8	5.96	0.89	−0.69	0.91	4.6	94.8	0.6
J_9	6.73	1.31	−0.36	0.68	12.40	86.42	1.44
J_{10}	6.01	0.69	−0.31	1.38	0.90	98.70	0.40
J_{11}	6.76	1.36	−0.48	0.76	1.36	94.44	4.20
J_{12}	6.18	1.38	−0.89	0.71	5.80	91.41	2.79
J_{13}	5.53	0.56	0.86	2.60	4.44	94.40	1.26
J_{14}	5.66	1.39	−0.79	0.78	0.19	96.13	3.68
J_{15}	6.58	1.09	−0.61	0.54	16.10	80.11	3.79
J_{16}	5.45	1.36	0.39	0.67	3.36	96.51	0.13
J_{17}	5.98	1.08	−0.56	0.64	2.30	96.50	1.20
J_{18}	6.01	0.59	−0.36	1.43	1.39	95.89	2.72
J_{19}	6.03	0.74	−0.09	1.07	5.47	84.41	10.12
J_{20}	6.33	0.77	0.35	1.17	29.46	61.35	9.24
J_{21}	5.76	1.38	−0.47	2.01	7.01	90.35	2.64
J_{22}	6.05	0.49	−0.28	1.87	4.61	86.87	8.51
J_{23}	6.41	0.88	0.45	2.17	27.04	67.63	5.33
J_{24}	6.38	1.03	−0.59	0.87	11.90	85.82	2.28

6.14 Texture of Marsh Sediments

Texturally the marsh sediments are composed of 90 % silt with sub equal proportions of fine sands and clays forming the rest (Table 6.4). The marsh sediments are mostly moderately well sorted to moderately sorted with graphic mean size belonging to the silt fraction. Most samples show a slight tendency of negative skewness. The negative skewness in the marshy region of the estuaries perhaps indicates trapping of larger bed-load particles by the marsh vegetations. The cumulative curves mostly show a high peakedness with greater K_G values. The statistical size parameters of the sediment samples are very much analogous to the lagoon or distal shelf sediments.

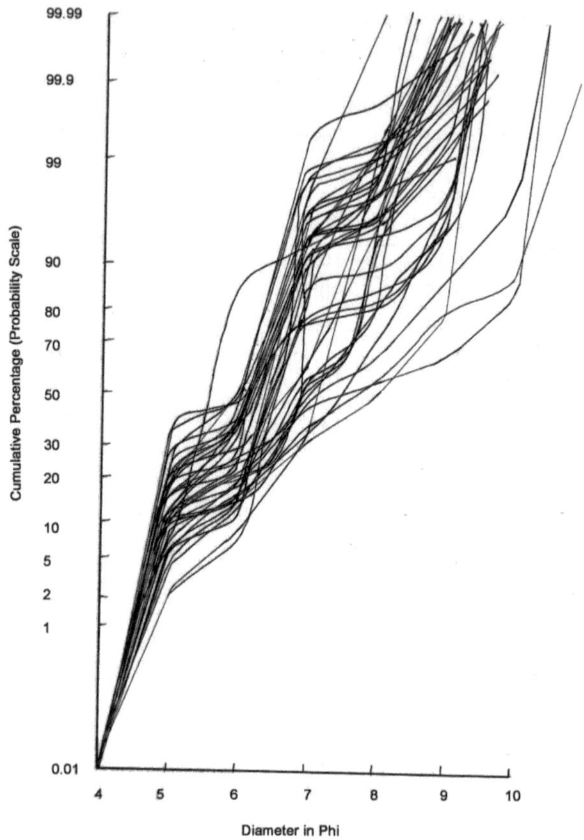

Fig. 6.6 Cumulative curves of the collected sandy mud sediment samples

Table 6.4 Textures of the marsh sediments

Sample no	Graphic mean (Mz)	Inclusive graphic standard deviation (σ_1)	Inclusive graphic skewness (SK_1)	Inclusive graphic kurtosis (K_G)	Sand %	Silt %	Clay %
B_4	6.05	0.84	−0.10	1.37	0.80	98.80	0.40
B_5	6.31	0.87	0.45	2.17	1.32	94.48	2.79
B_6	6.76	1.40	−0.37	2.08	6.89	90.32	1.12
B_7	6.50	0.59	−0.18	1.93	3.44	95.44	3.72
B_8	5.41	0.78	0.35	3.17	0.16	96.12	3.76
B_9	5.46	0.19	−0.69	0.79	18.12	78.12	3.76
B_{10}	7.76	1.35	−0.38	0.78	2.36	97.53	0.11
B_{11}	7.01	0.61	−0.41	1.53	1.36	94.88	3.76

6.15 Texture of Mangrove Swamp Sediments

The tidal flats are mostly siliciclastic. The sediments are generally silty to sandy. All the cumulative curves are non-lognormal. The silty sediments are extremely zig-zag and reflect deposition out of ebb-flood cycles. The statistical size parameters shows that the sediments are mostly well sorted to moderately sorted with graphic mean size belonging to the silt fraction (Table 6.5). Most sediment samples show a slight tendency of negative skewness. The cumulative curves mostly show a high peakedness with K_G values often greater than 1.0. The cumulative curves for the sandy sediments show three distinct sub-populations viz. rolling, saltation and suspension sub-populations. The cumulative curves for the silty sediments, however, show more intricate zigzag patterns reflecting ebb-flood cycles.

6.16 Texture of River Sediments

Collected sediment samples from different locations of the rivers were analysed. Texturally the sediments are 90 % silts with subequal proportion of the fine sand and clay. Statistical size parameters were calculated using formula of Folk and Ward (1957). Graphic mean size (Mz) 6.74 phi indicates low competency of the river (Table 6.5). Sediments are well sorted to moderately sorted (σ_1) ranging from 0.31 to 1.4 phi. Most grain-size distribution curves reveal a tendency of slightly negative skewness (Das 2009). A general seaward increase of grain size along the river valley indicated removal of fine sands because of higher tidal amplitude and wave vigour near the mouth.

Table 6.5 Statistical grain size parameters of the mangrove sediment samples

Sample no	Graphic mean size (Mz)	Inclusive graphic standard deviation (σ_1)	Inclusive graphic skewness (SK_1)	Inclusive graphic kurtosis (K_G)	Sand %	Silt %	Clay %
S_1	6.97	0.99	−0.41	1.68	3.75	92.30	4.05
S_2	6.89	0.76	−0.33	0.78	1.24	97.13	1.73
S_3	6.86	1.40	0.46	0.91	0.94	94.47	4.79
S_4	6.76	0.64	−0.48	0.84	6.43	93.01	0.56
S_5	6.81	0.43	0.64	0.43	18.16	81.43	0.41
S_6	5.99	0.81	−0.36	0.67	3.18	88.94	7.88
S_7	6.13	0.49	0.31	0.92	14.12	84.01	1.87
S_8	6.31	0.93	0.39	0.78	7.17	89.31	3.52
S_9	5.98	0.59	0.52	0.97	8.29	88.11	3.60
S_{10}	6.37	0.62	0.11	0.93	7.15	89.99	2.86
S_{11}	5.92	0.48	−0.33	0.78	2.91	94.32	2.77
S_{12}	6.77	0.41	−0.037	0.67	9.13	87.19	3.68

6.17 Textural Sensitiveness

There is rhythmicity in the nature of deposition, which perhaps indicates the depositional pulses for the ebb and flood flows through the rivers. The inflection points in the curves between successive ebb flood cycles are marked at 2.5–3.0 phi, 3.5 phi, 5–6 phi and 8 phi sizes respectively. Such rhythmicity in the nature of depositional behaviour of tidal sediments is supposed to be highly process – responsive.

Sandy sediments (ranging from 1.5 to 4 phi) from the point bars and mid channel bars of estuaries show a complete different pattern from that of sandy silt sediments. The cumulative curves for these sediments show a non-linear pattern and are nicely comparable to that of the other point bar sediments. These sediments show two major inflections one at 2.25 phi and the other at 3.0–3.5 phi. These inflections divide the curves into three sub populations as rolling, saltation and suspension respectively. The saltation population generally constitutes about 75 % of the total material, the rest being deposited by either rolling or suspension. Silty sediments having range from 4 to 9 phi from similar areas also show exactly the same pattern where the three above mentioned subpopulations can be well recognised. In these curves, the inflections take place at 6–7 phi respectively. The central saltation population constitutes about 70 %. These sediments do reflect their deposition from tractive movements of water in a unidirectional flow condition.

6.18 Summary

As a matter of fact grain size of the estuarine sediments reflects the nature of source sediments and their hydrodynamic condition of deposition. Generally erosion prevails towards the seaward reach of the estuary with high wave energy and deposition dominates in the landward reaches of relatively quieter environment. Thus finer muddy sediments are deposited on the estuarine banks, flanks of the mid channel bars and point bars with low depositional energy.

Sediment grain size is considered as one of the important tools for the interpretation of depositional environment in the estuaries (Das 2009). Sorting indicates the process of modifications of the sediments, whereas graphic mean size reflects the environment of sediment accumulation. Rigorous flow transports the sediments in the depositional environment causes poor sorting of sediment particles. Sediments are skewed in selective transportation and a particular sediment population is characterised by inclusive graphic kurtosis (Davis 1983).

The cumulative curves are mostly non-lognormal with rolling, saltation and suspension populations (Visher 1969) as recorded from most of the samples collected from the different geomorphic situations. The saltation and suspension sub-populations are rather undifferentiated in sandy samples of both beach and mid channel bars. The statistical size parameters and the points of inflections in the cumulative curves for the mangrove sediments samples as shown in the tables and figures are the indicators of depositional environments for the Sunderbans. The inflections points between the rolling and saltation transport for the sandy samples occurs at a finer size than that for the muddy sediment samples. This indicates that the agent responsible to bring about a change in the mode of transport of particles was weaker for the sandy samples compared that of the muddy samples.

References

Benson DJ (1981) Textural analyses with Texus Instruments 59 programmable calculator. J Sediment Petrol 51(2):61–62

Blott SJ, Pye K (2001) GRADISTAT: a grain size distribution and statistics package for the analysis of unconsolidated sediments. Earth Surf Process Landf 26:1237–1248

Coulter Electronics, Inc. (1989) Coulter Multisizer AccuComp color software: reference manual. Hialeah

Das GK (2009) Grain size analysis of some beach sands from the Indian Coasts. Geogr Rev India 71(1):10–18

Davis RAD (1983) Depositional systems. Prentice-Hall, INC., Englewood Cliffs

Folk RL, Ward W (1957) Brazos River bar – a study of the significance of grain size parameters. J Sediment Petrol 27:3–26

Fuller AO (1961) Size characteristics of shallow marine sands from Cape of Good Hope, South Africa. J Sediment Petrol 31:256–261

Jones SB, Simpkin P (1970) A computer program for the calculation of hydrometer size analyses. Mar Geol 9:23–29

Kane WT, Hubert JF (1962) FORTRAN program for the calculation of grain-size textural parameters on the IBM 1620 computer. Sedimentology 2:87–90

Krumbien PD (1938) Size frequency distributions and the normal phi curve. J Sediment Petrol 8:84–90

Krumbein WC, Pettijhon FJ (1938) Manual of sedimentary petrography. D. Appleton – Century, New York

Muerdter DR, Dauphin JP, Steele G (1981) An interactive computerized system for grain size analysis of silt using electro-resistance. J Sediment Petrol 51:647–650

Poppe LJ, Eliason AH, Fredericks JJ (1985) APSAS: an automated particle-size analysis system. U.S. Geological Survey Circular 963

Rigler JK, Collins MB, Williams SJ (1981) A high – precision digital recording sedimentation tower for sands. J Sediment Petrol 51:642–644

Sawyer MB (1977) Computer program for the calculation of grain-size statistics by the method of moments. U.S. Geological Survey Open File Report, pp 77–580

Schlee J (1966) A modified Woods Hole rapid sediment analyzer. J Sediment Petrol 30:403–413

Schlee J, Webster J (1967) A computer program for grain-size data. Sedimentology 8:45–54

Spencer DW (1963) The interpretation of grain size distribution curves of clastic sediments. J Sediment Petrol 33:180–190

Syvitski JPM, Asprey KW, Clattenburg DA (1991) Principles, design and calibration of settling tubes. In: Syvitski JPM (ed) Principles, methods and applications of particle size analysis. Cambridge University Press, New York, pp 3–21

Visher GS (1969) Grain size distributions and depositional processes. J Sediment Petrol 39(3):1074–1106

Zeigler JM, Whitney GG Jr, Hayes CR (1960) Woods Hole rapid sediment analyzer. J Sediment Petrol 30:490–495

Chapter 7
Estuarine and Coastal Erosions

Abstract Sunderbans is characterised by anastomising network of tidal rivers, creeks and minor salt water courses. This estuarine coast is undergoing continuous changes due to the action of the asymmetric tidal flow regime, river fluxes, waves, sea level change and the vagaries of nature in the form of cyclones and storms which are almost regular phenomena. Depositional and erosional features are observed at river banks, point bars, mid channel bars and the intertidal swash platforms. Sandy deposits are eroded because of resuspension and transported by seaward waves and tides whereas mud is eroded by scouring and transported by suspension. The erosion and accretion process over the Sunderbans delta might have a bearing on the neotectonic activity on the Bengal Basin.

Keywords Erosion • Accretion • Flood and ebb flow • Bank subsidence • Sediment trapping • Slumping • Undercutting

Sunderbans – the largest prograding delta of the world along the coastal Bay of Bengal constitute such tidal river complex, which has given rise to a number of features towards elucidation of the depositional and erosional behaviour of the delta complex in space and time. The process of erosion and deposition of the meandering tidal rivers is primarily controlled by transverse helicoidal flows coupled with the landward and seaward movement of flood and ebb currents acting along the course of the rivers. The entire Sunderbans area is intersected by a large number of tidal rivers estuaries, tidal creeks and minor salt-water courses. This network of drainage with dynamic flow patterns of 31 tidal rivers in Sunderbans regions, the process of erosion make the geomorphic set up of this region a complex one, when studied from north to south as well as from east to west.

The process of erosion of the estuarine Sunderbans is primarily controlled by tidal flows coupled with the landward and seaward movement of flood and ebb currents acting along the estuaries. Flood and ebb flow follow different sides indicating the existence of both flood-dominated and ebb-dominated sides in this estuary along with velocity differences in the flood and ebb flows. Generally tidal asymmetry governs the long-term trend of erosion on the banks and mouth regions of the estuaries. Many of these serve dual purposes of fresh water discharge from upland and to and fro movement of coastal brackish water from the seaward

G.K. Das, *Estuarine Morphodynamics of the Sunderbans*,
Coastal Research Library 11, DOI 10.1007/978-3-319-11343-2_7

direction. The erosional features are seen generally at estuarine river banks, point bars of marginal and crescentic types, mid channel bar, mouth bar or the intertidal swash platforms. The network of drainage with dynamic flow patterns of estuarine regions, the process of erosion make the geomorphic set up of the estuary a complex one. Rivers form funnel shaped estuaries at their mouths being the width of about a few to tens of km in average. In the funnel shaped estuaries, the tidal amplitude will increase significantly. As a result, tidal water volume increases and alters estuarine dynamics. Much of the estuarine plain are inundated by floodwater during wet seasons but during dry seasons the estuarine plains are completely dried up and even register evidences of desiccation with salt encrustation (Das 2008, 2009b).

7.1 Controlling Factors for Erosions

Larger channels like Saptamukhi, Matla, Thakuran, Bidya in the estuarine Sunderbans, flood and ebb flow follow different sides of the tidal channels. This suggests that there is a flood-dominated side and an ebb-dominated one within the same tidal channel. Further, asymmetry in the direction of flood tide results the erosion along the tidal channel in the estuarine region and the duration of asymmetry governs the trends of erosion in the channel system (Ginsberg and Perillo 2004). The neap-spring cycles, the flood-ebb cycles and the tidal regime within the estuaries have influenced the configuration of the marginal levees, accretionary point bars, mid channel bars and river mouth bars. The process of accretion in the estuarine Sunderbans is mostly controlled by a higher rate of sediment trapping in areas of geomorphic highs covered with network of mangrove vegetation. Subsidence of the bank with simultaneous sediment filling all along the tidal courses is quite common. Sediment is also baffled by the salt marshes on the banks and newly emerged islands of estuaries.

Thirty one rivers forming a network of intricate minor salt-water courses intersect the Indian Sunderbans. These rivers and tidal inlets serve dual purposes of freshwater discharge and to and fro movement of saline water wedges arising from the sea face. The coastline of West Bengal is comprised of the tidal plains of a large number of estuaries and tidal inlets of which the Ganges-Brahmaputra tidal plain is the most important. Tidal flooding, river fluxes, waves, sea level change, regular episodic and non episodic events like cyclones and storms are the dominant physical processes that continually modify this coast in a short and long term basis (Das 2009a, b).

7.2 Flow Pattern

Every meandering drainage system with its erosional and depositional sites are the resultant products of two distinct flow pattern (i) the downstream sinuous flow pattern and (ii) Helicoidal flow pattern. The mid channel bar in a meandering system is either the result of slackening flow velocity due to channel topography or it is a result of isolation of the point bars due to subsequent cut off by the later deflection of the water flow. In a tidal river the depositional behaviours of the channels are further very much sensitive to the seasonal variation of the tidal behaviour. The mid channel bars in a tidal river are mostly resultant from convergence of flood and ebb flows (Fig. 7.1).

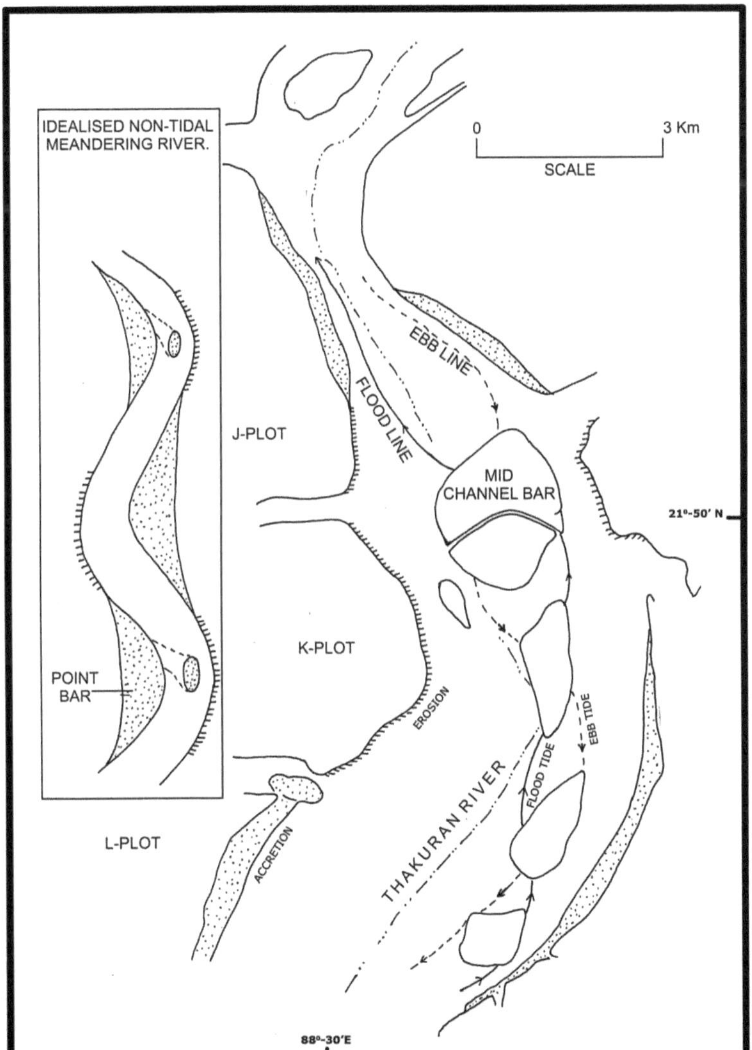

Fig. 7.1 Formation of mid channel bars resulting after convergence of flood and ebb flows

7.3 Geomorphic Sites of Erosion

Depositional and erosional behaviours along convex and concave banks of the meandering stretch of the rivers of the longest prograding delta, Sunderbans are recognizable. All the depositional and erosional features are seen generally at riverbanks, point bars of marginal and crescentic types, mid channel bar, mouth bar or the intertidal swash platforms.

Rivers of Sunderbans are highly seasonal and tidal flowing towards Bay of Bengal and form funnel shaped estuaries at their mouths having width of about 6–8 km in average. In the funnel shaped estuaries, the tidal amplitude will increase significantly. As a result, tidal water volume increases and alters estuarine dynamics. During the dry season the river regimes are controlled by tidal water, whereas during rains the headwater discharges have strong influence over river regime. Holocene coastal and estuarine plains flank the riverbeds. Much of the estuarine plain of Sunderbans are inundated by floodwater during wet seasons but during dry seasons the estuarine plains are completely dried up and even register evidences of desiccation with salt encrustation. During dry period, the tidal flows are accumulated in the river channels. Discontinuous stretches of mangrove forests fringe the river. The flood plains are almost horizontal and are composed of fresh water clays (Das 2010a).

7.4 Characteristics of Estuarine Banks

Estuarine banks generally are the sites of erosion. In places of massive erosion, the banks have been fortified by construction of embankments. Landward, the intertidal area behind the embankments is occupied by marsh vegetation (Bhattacharya 1999). The topmost portions of the bank deposits of the estuaries register evidences of long exposures and contain desiccation cracks and salt-encrusted surfaces in patches. The following features characterize the supratidal levee deposits – marshy areas are characterized by marsh lamination with fine obscured laminae of silt and clay profusely disturbed by roots of marsh vegetation. Rapid thinning of sand laminae and dominance of mud in the upward direction bank deposits are evident in many estuarine bank sections. Plant remains and plant roots constitute a substantial component of bank deposits, particularly towards their upper portions. Levee surface sometimes shows salt-encrusted patches due to evaporation of saline water pools formed during over bank flooding. Suspension mud deposits on the topmost part of the river banks exhibit mud cracks or curly mud cracks and indicates sub- aerial exposure. Curly mud cracks are often associated with areas having algal matting (Das 2010b).

7.5 Characteristics of River Banks

Both banks of the rivers of Sunderbans are generally characterized by fringing mangroves. As a result, roots and pneumatophores of mangroves and often heavily disturb the nature of stratifications of the bank deposits by the churning effect of burrowing organisms. In places of massive erosion, the banks have been fortified by construction of embankments. As a result mangroves are truncated in the intertidal zone. Landward, the intertidal area behind the embankments is occupied by marsh vegetation (Bhattacharya 1999). The topmost portions of the bank deposits register evidences of long exposures and contain desiccation cracks and salt-encrusted surfaces in patches. The following features characterize the supratidal levee deposits: Marshy areas are characterized by marsh lamination with fine obscured laminae of silt and clay profusely disturbed by roots of marsh vegetation; Rapid thinning of sand laminae and dominance of mud in the upward direction bank deposits are evident in many bank sections; Plant remains and plant roots constitute a substantial component of bank deposits, particularly towards their upper portions; Levee surface sometimes shows salt-encrusted patches due to evaporation of saline water pools formed during over bank flooding and suspension mud deposits on the topmost part of the riverbanks exhibit mud cracks or curly mud cracks and indicates sub aerial exposure. Curly mud cracks are often associated with areas having algal matting.

In vertical sections the bank deposits are separable into three units based on colour from the top to the mean low water line. The bank top is dominated by mangrove species *Phoenix paludosa*, *Ceriops decandra* and *Excoecaria agallocha*. The top stratum mud deposits (up to 2 m thickness) are rigorously disturbed by the penetration of mangrove roots up to a depth of 1 m or so. The immediately underlying unit (generally of 1.5–2 m thickness) is profusely bioturbated by crab burrows and penetration of mangrove rootlets. The muddy burrow walls and root affected mud cliffs display clear evidences of oxidation marked by brown colour of sediments. The lowermost unit of the banks (up to 2 m thickness) is less affected by bioturbational disturbance. In majority of the cases, the topmost unit has an erosional contact with the underlying intermediate unit, whereas no such erosional contact is recognizable between the intermediate and the lowermost units excepting a distinct colour variation.

7.6 Erosional Processes and Features

The estuarine banks register evidences of local sagging (subsidence) in many instances (Fig. 7.2). These areas of subsidence appear as troughs along the banks to be filled up partially or completely with newly accreting sediments. On the initial stage of compaction these sediments support plant saplings. In addition, collapsing of banks is locally common due to over steepening. Interbedding of sands with

Fig. 7.2 River bank erosion through sagging-down of bank materials

mudclast enriched layers is a common feature. Eventually, some portions of the upper levels of the bank, slump down to the lower levels. This leads to a mixing of different stratigraphic levels. The over steepening of the banks is promoted by loosening of bank materials which interalia is caused by the following important factors (i) Semi-diurnal tidal fluctuations and (ii) hydrodynamic conditions that operate in meandering rivers in the form of helical flows. Somewhere, bank erosion takes place in a piecemeal manner, which adds to the processes of lateral shifting of the estuaries leads to the piecemeal mechanism of bank erosion.

Data collected along and across the rivers of Sunderbans since 1989 reveal four different mechanisms of bank erosion. These are: (i) erosion due to impact of longitudinal and helicoidal flow in the meandering stream (ii) the thrust of the ebb and flood cycles and fluctuation of the high tidal amplitude (iii) cohesion of grains attributed by textural parameters e.g. banks composed of sands are eroded more than these of clay or more cohesive material and (iv) differential subsidence causing piecemeal bank erosion and later modification by waves near the mouth of the rivers.

7.6.1 Undercutting and Collapsing

Undercutting of the mangroves is well documented and in the river banks by the hanging of (i) root system of mangroves at low tide and (ii) exposures of mangrove supporting soil chunks at low tide. At high water stage, the tidal water intrudes below these chunks of topsoil and ultimately causes collapsing of roofs. These subsided areas are simultaneously filled up with tidal sediments and became parts of the intertidal zones very quickly. The areas act as newly accreting zone where saplings of mangroves come to constitute the elderly mangroves in their order of appearances (Das and Bhattacharya 1994). Occasionally, the stable river banks with an area of some tens of square meters collapse enblock together with mangrove trees.

7.6.2 *Subsidence and Rapid Sedimentation*

Evidences of subsidence and rapid rate of sedimentation in areas of collapsing are marked by (i) Erosion of riverbanks (ii) Undercutting of mangrove forests (iii) Exposures of chunks of mangroves and (iv) Supratidal bluff breccias in the intertidal stratigraphy. The banks of the rivers of Sunderbans register evidences of sagging down in certain portions (Fig. 7.2) in alternation with areas of a gentler slope to the order of 3°–5°. For eg, in one area of the left bank of river of Thakuran subsidence of the banks to the order of 15 m has been recorded in 1982. Subsequently during last 10 years the entire subsided region has been filled up to be exposed as intertidal zone measuring an average of 1.5 m annual rate of sedimentation. The terraced bank pattern of rivers in the Sunderbans can be supposed to be the result of erosion followed by differential subsidence and filling up of sediments.

7.6.3 *Piecemeal Mechanism*

The river banks of the estuaries show evidences of collapsing in certain portion along the length of the estuaries. Stretches of the estuarine banks in between collapsed zones exhibit a gradual slope of 3°–5° towards the estuarine side. Collapsing of banks involves destruction of the older vegetations, which are buried simultaneous with bank erosion. In estuarine bank sections, the plants often lie with upright tree trunks. Occasionally tilting or flattening of the trunks is noticed off and on.

Near the mouth of the estuaries certain transverse mud bars are seen to be exposed in low tides. These bars very commonly show a direct connection with the banks and raise their heads above the low tide level. They have a general tendency of tapering away from the bank. The exposed portions are often segmented by the passage of very small creeks, the alignments of which are more or less parallel to the direction of tidal currents.

Somewhere relict mud bars in the form of small isolated patches are present within the marginal waters. This relict bars are remnants of the transverse mud bars which retain their existence after removal of certain portion of the transverse bars. The tidal currents flowing through the small creeks generally initiate removal of materials. These observations on the presence of transverse mud bars and relict bars within the marginal waters of the estuaries suggests that the process of estuarine migration, in addition to the usual process of erosion and deposition in unidirectional flows, is further enhanced by the piecemeal mechanism of bank erosion (Fig. 7.3).

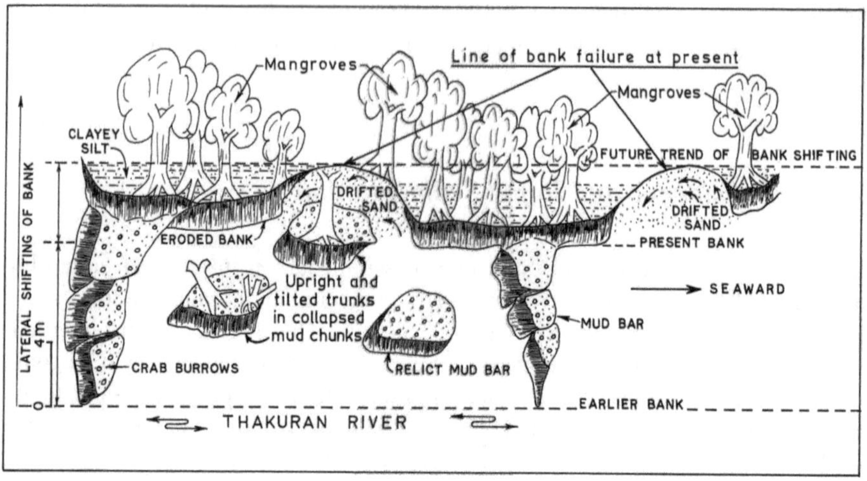

Fig. 7.3 Piecemeal mechanism of bank erosion

The stretches of the banks in between the transverse mud bars exhibit collapsing and refilling by sandy sediments forming slopes of 3°–5° towards the estuaries. The washing of the mud banks often give to flow of muddy slur over the sandy sediments, thus giving rise to a sharp contrast between the lower sandy sediments and the upper muddy sediments (Das 2008).

7.6.4 Furrow and Ridge Structures

A few erosional features like furrow and ridge structure, mud mounds and armoured mud balls are present in the places of severe erosion along the coast near the mouth of the estuaries. The crestal alignment of these furrows indicates their ebb current origin supported by the impacts of swash and backwash. Rippled or smooth structured sandy furrows are originated in a lower to transitional flow regime. They occur within the mud ridges with a sand-river appearance when these furrows extend in a sinuous train for several meters. Strong ebbing currents carry the large oyster shells and pieces of boulders which are scattered all over the furrows. The alignment of these ripple crests is at right angles to the ridge axes. Despite adequate cohesion in mud, the mud flat is cut into several sub-parallel furrows in the areas of severe erosion. These structures appear like whale back and run discontinuously dissecting the mud flat. The furrow and ridge features are considered as evidences of estuarine erosion having ebb tidal origin, although these structures are modified by wave erosion in some occasions (Bhattacharya 2000).

7.6.5 *Mud – Mounds*

The mud mounds occur in the sub-aerial environment of the coastal region with a space of tens of meters in many occasions with elliptical to sub-elliptical shapes. These mounds with horny projections are generated on the zones of severe erosion. These features having the scours or grooves at the central parts with low height are surrounded by rippled sand flats. The landward directions of the bifurcated horns of the mud mounds indicate their origin from corrosion of backwash flows. These features are rather scarce in the modern and ancient records (Allen 1984) along the coastal tract in the estuarine environment.

7.6.6 *Mud Balls*

Mud balls, derived from the mud chunks that collapsed from the mud banks due to high tidal vigour and wave actions are strewn over the stretch at the transitional zone of the coastal region. The mud balls attain different shapes like cylindrical, spindle, ellipsoid, sphere etc. (Bell 1940) before becoming mud balls through the rolling mechanism caused by the inundation twice daily in the semidiurnal tidal situations. Molluscan shells either intact or broken (Gastropods and Pelecypods) are found in these mud balls which are either within the mud lumps earlier or inserted later from the beach during rolling by the tidal as well as wave actions. These mud balls, compositionally made up of silty clay are covered with a few fine sand grains and biotic materials added during rolling. Occurrences of these mud balls (Fig. 7.4) of

Fig. 7.4 Mud ball creates as an obstacle to form the current crescent at intertidal region. The pen is 15 cm long

mechanical origin throw some light upon erosion of the coastal environment that helps in the interpretation of hydrodynamic environment of the estuaries.

7.7 Erosion at Estuarine Mouths

Significant reconfiguration of the estuaries is the resultant effects of erosion at its distal end part near the coastal tract. Erosion in the mouth of estuaries generally occurs in the tectonically uplift and well developed cliff areas with the beating of waves against the shore. The elevated portions are continued to erode as long as the land mass becomes parallel to that of the sea surface. Waves, currents, tides and winds continuously shift the sediments in the coastal region between the estuaries and seas. Erosion in the coastal areas surrounding the mouth of the estuaries is resulted by the various geological and tectonic factors along with tidal range, currents, waves and volume of flow, speed and direction of winds, effects of storm surges and sea level fluctuations. The erosional coastal region is generally ornamented with the erosional sedimentary structures like mud mounds, mud balls, erosional furrows and ridges etc. Coastal erosion leads to the land ward displacement of the shoreline.

7.8 Coastal Erosion

Recent data show an alarming rate of coastal erosion in Sunderbans as well as the trend of sea level rise in that region. Erosion is an active process all along the shorelines of the Sunderbans. The coastal shoreline of the Sunderbans – the buffer zone between the land and sea is unstable as the waves, currents and wind continuously shift the sands (Uehara et al. 2010). This area is considered as highly erosive one. Erosional structures as obstacle scours are frequently found in upper and middle foreshores of the sandy coast. Large clasts such as pebble and gravel size clay balls, fragments of woods occur in the sandy sea beach of river mouth bars. These clasts create obstruction on the surface when the stream flows during the back wash of waves and form scour troughs during strong tidal surge. All these features coupled with the available biogenic activities may have a short term and long term effects on the changing configuration of the tidal flat surfaces and shore line of Sunderbans coast. The northeastern, southeastern and southwestern sides of Sagar Island are facing vigorous erosion due to the concerted acts of various natural processes and anthropogenic activities. The erosion rate from 1996 to 1999 was calculated to be 5.47 m/year (Gopinath and Seralathan 2005).

The coastal area of Sunderbans has a gentle seaward slope and major portion of it lies between 7 and 8 m above mean sea level. The meso-macrotidal regime inundates most areas twice daily. Erosion in the coastal areas is led by gradual denudation as well as by catastrophic events. Gradual erosion results from (i) flood and ebb tidal currents (ii) high tidal range and (iii) strong littoral drift. Major catastrophic causes include (i) cyclonic storms and hurricanes (ii) storms – induced waves (iii) tidal bores and (iv) tectonic (neotectonic) movements (Das and Bhattacharya 1994).

7.9 Mechanisms of Coastal Erosion

Two types of mechanisms are found responsible for the coastal erosion of the deltaic Sunderbans:

(i) Deposits of sands are worn because of resuspension and are transferred seaward by waves and tides.
(ii) Mud is eroded by scouring and carrying away to the subtidal region by suspension transport.

Process of erosion is accelerated by the exotic boulders, dead shells and large sand particles. Those act as tools to augment the process of erosion.

7.10 Biological Evidences of Erosion

Sunderbans is a biogenous coast where biological factors play a significant role in coastal evolution. Activities of mangrove fauna in the intertidal flats and coastal regions develop micro—morphological features which are very much significant for the trapping of sediments and mangroves seeds as well as mark the evidences of the erosional features and are sometimes helpful on computing the loss of sediments due to erosion.

There is always a conflict between the biogenic and physical processes in a beach. The physical processes are responsible to generate sedimentary structures of a beach whereas, the various benthic organic communities, in the contrary, try to obliterate and modify the original sedimentary structures. Biota, biogenic activities, bioturbation structures etc. are of late suggested as players for modifying the sedimentation pattern, beach morphodynamics and even as biological tool to determine amount of erosion (Roychaudhuri et al. 2003) of the intertidal beach environment. A gastropod mollusk, *Amalda ampla* (Gmelin) forms conical dome like structures (Fig. 7.5) as their resting traces at the intertidal beach areas of Frasergunj- Bakkhali along the coastal tract of West Bengal, India. A typical saw like indentation mark is created after cleavage of the conical dome like bioturbation structure demarcated as grazing traces (Fig. 7.6) when this mollusk is in the movement for searching of food. The trail of their feeding tracks in the form of linear zigzag depression is well preserved on the fine sediment surface.

Fig. 7.5 Conical dome like bioturbation structure by *Amalda ampla* over the interference ripple marks in the intertidal zone. The pen is 15 cm long

Fig. 7.6 Indented pattern of grazing traces of *Amalda ampla* on the intertidal zone after recession of water. The pen is 15 cm long

7.11 *Amalda ampla* – A Biological Indicator of Erosion

Amalda ampla-a gastropod mollusk mostly occurred in the intertidal sands of erosional sites covered with thin veneer of silt may be considered as a very useful biological indicator of beach character. *Amalda ampla* occupies well- defined zones of fine sandy substrates with higher concentration of metal and metal oxides. This gastropod prefers coastal waters with less hardness that inundates the intertidal zone during high tide.

Frasergunj –Bakkhali lying in between Latitude 21°35′N and Longitude 88°15′E along the coastal tract of West Bengal, Eastern India-a high revenue-earning sea beach for tourism industry belongs to the hostile meso-macro tidal realm. Morphodynamics of this beach of 1.2 km stretch between Frasergunj at the west and Bakkhali in the east is controlled by sea waves modified by the action of tide and wind as the principal architects. This intertidal zone shows the depth ranges from 1 to 5 m. The intertidal areas of this beach registers erosion resulting from flood and ebb tidal currents, high tidal range and strong littoral drift including catastrophic causes like tidal bores, storm surges, cyclonic storms due to atmospheric depressions and neo-tectonism (Das and Bhattacharya 1994). *Amalda ampla* prefers this erosional zone of intertidal areas of Frasergunj –Bakkhali as their habitat.

7.11.1 Characteristic Features of *Amalda ampla*

Amalda ampla of the Family-Olividae; Order-Neogastropoda; Class-Gastropoda under the Phylum-Mollusca is common on fine sandy beaches from intertidal zones to 4 m depth with sometimes occurring buried in the tidal shoals along the coastal stretch of Frasergunj –Bakkhali although this gastropod is usually an offshore form. Slender elongated shell of medium size that often attains up to 30 mm in length is highly polished, ivory-white to creamy –brown in colour. Melanin and pyrroles are the pigments involved in colouring (Marshall and Williams 1982). Carnivorous *Amalda ampla* feeds mostly on living and dead tissues of bivalves, crustaceans and other invertebrates (Kilburn 1977).

Dwelling as well as feeding structures made by *Amalda ampla* have profound role in churning the intertidal fine silty sand substrates. An overall mixing of sands with the thin veneer of mud takes place very commonly and this alters the textural properties of sediments deposited by physical processes. The primary depositional behaviour of sediments produced by swash – backwash (waves) effects and by tidal currents is manipulated largely by different intensities of biogenic activity of *Amalda ampla*. Net sediment budget of sand and mudflats are remaining constant; burrowing leads to both lateral and vertical exchange of sediments and biogenic subduction of surface topography. The internal laminations vertically up to 5 cm are disturbed to different degrees and are deformed by folding and microfaulting. The ejected burrow materials in the form of mounds commonly create complications in the surface flow pattern and flow directions. Flow divergence against their mounds and obstacles becomes responsible for the generation of current crescents and modifications of bed forms. *Amalda ampla* may be considered as biological indicator for the determination of the beach erosion as observed at Bakkhali for the period from 1986 to 2011. *Amalda ampla* abundantly occurred only during the erosional phases of sea beaches at Bakkhali and Frasergunj respectively.

7.11.2 Physico – chemical Parameters

Physico-chemical parameters of both coastal waters and beach sediments along with soil textures are very much important for the occurrences of *Amalda ampla* in this particular habitat. Analyses of water samples collected during high tide at Frasergunj-Bakkhali shows higher suspended sediment concentration (62–130 mg/L) and comparatively lower hardness (1,231–1,305 ppm) of coastal waters. The concentration of dissolved trace metals (Cu, Cd, Fe, Mn, Zn, Pb, Ni and Co), taking the higher values in all the water samples analyzed, is 162 ppb. The Dissolved Oxygen (DO) content also shows variations of values ranging from 6.6 to 8.1 mg/L. The value of DO is high in the near shore waters due to the occurrence of maximum turbulence in comparison to offshore areas. Salinity is variable that ranges from 14 to 31 ppt annually as compared to the water temperature 18–35 °C. The pH values remained constant (8.1–8.3) except during the monsoon months (July to October), when it comes down to minimum 7.5–7.7.

7.11.3 Sediment Texture

Texturally, the sediments are generally composed of almost equal proportion of fine sands and silts. The substrate characters of the habitat of *Amalda ampla* consist of moderately to well-sorted sediments having 48 % sand, 43 % silt and 9 % clay. Copper varies from 10 to 100 ppm, while lead and zinc varies from 5 to 30 ppm and from 15 to 25 ppm respectively as heavy metals. In general, the higher values of heavy metals are restricted in the finer fractions like sandy silt and/or silt. This fine sand flats are generally with less surface structures and covered with the veneer of mud accreted through suspension.

7.11.4 Habitat Structure of *Amalda ampla*

Apart from some very basic erosional structures of this intertidal beach environment mudballs, furrow and ridge structure, fragment of wood trunk as obstacle scours, the presence of *Amalda ampla* also suggests the erosion as biological indicator.

The Frasergunj-Bakkhali shoreline is a highly erosive one and *Amalda ampla* are available only at the site of erosion with fine sediments of sand and silt. This particular habitat of *Amalda ampla* in the intertidal areas of Frasergunj –Bakkhali is characterized by (i) the substrate quality of beach sediments composed of more or less equal proportions of sands and silts, (ii) coastal waters with maximum turbulence, (iii) high concentration of trace metals and metal oxides, (iv) high DO content in the coastal waters, (v) intertidal zone with severe erosion, (vi) maximum sediment accreted through suspension, (vii) rich in benthic fauna like crustaceans, polychaetes etc. as they feed on them.

7.12 Estimation of Beach Erosion Using Biological Tools

Biogenic activities sometimes may be helpful for the determination of the rate and nature of beach erosion. The study was carried out on a portion of a thin veneer of loose sands of only 9 cm layer over the sandy intertidal area of Frasergunj (Lat. 21° 34′ N and 88° 15′ E), India. The selected study area of 178 m length and 41 m width was characterized with the loose sands which were accumulated probably due to the accretion with the wave borne sands from the Bay of Bengal. This zone of loose sands, very rich in organic matter was the habitat for the *Ocypode macrocera*, commonly known as Red crab. The study was carried out during the period between 2006 and 2011.

Red crabs under phylum Arthopoda are very much selective of dwelling places in their burrows. It is observed that they never enter into other burrows, when they are caught and released in the separate places far away from their own burrows. Normally the burrows are cylindrical in shape, sometimes branched in depth, 15–20 cm long, vertical or diagonally inclined. Sometimes these red crabs heavily disturb the transitional zone of supratidal and upper intertidal beach areas with the ejected materials in such a manner that numerical count of the crab burrows are simply impossible. When this type of zone is heavily populated by the red crabs, the area shows depositional environment. But after a few months sediment lost from the environment may lead to the conversion of an elevated zone into a plain or a depressed zone. For this reason the presence of red crabs may be considered as a biological indicator of sedimentary processes from the particular zone of their habitat.

7.12.1 Physico-chemical Environment of Sediments

Coastal waters with salinity ranging between 15 and 31 ppt as compared to the water temperature 20–30 °C and pH ranging in between 8.0 and 8.3 inundate the habitat of red crabs twice daily. The sediments are well sorted and are generally siliciclastic in nature. Concentration of metals of Cu (30 ppm), Pb (85 ppm), Zn

(65 ppm), Ni (40 ppm), Co (40 ppm), Cd (5 ppm) in the sediments appear to be free from metal pollution. However, the highest concentration of Mn (5,000 ppm) is recorded at the habitat of red crabs, although Cr (125 ppm) value shows below the contamination level. The concentration of metallic oxides such as MgO (1.65 %), Na_2O (0.21 %) and K_2O (0.36 %) are very poor except Fe_2O_3 (17.81 %). Total organic matter (0.49 %) in these loose sands is rather alarming for the tropic level of the red crabs which are generally benthic feeders. Almost sandy beach sediments with such physic-chemical environment are rather befitted with the habitat of the red crabs.

7.12.2 Observations

The beach sands of the upper intertidal region heavily disturbed by these crab-made burrows were drastically undergone erosion when the ejected materials were washed away. When inundated these cylindrical burrows are almost filled up with the ejected sands and the wall around the burrow mouth is collapsed and looked like a half of a sphere after recession of water where the diameter of the burrow mouth increased to 6 cm after modification.

It is observed that the crabs sometimes take rest inside the burrow when the area is already watered in the flood tide. The red crabs have the book gills in order to assist their respiration in the occasional periodic water logged environment. But coming out of burrow in dry session they dig the separate burrow according to their own will. This causes the enhancement of rate of sediment loss vigorously.

It is observed that a huge amount of sediment of a 9 cm thick landmass having volume of 656.82 m^3 in an estimated area of 7,298 m^2 had been eroded drastically within a period of 57 months for the bioturbational activities of the crabs. The rate of erosion may be estimated within the beach area using this erosional pattern as biological tools. Applications of such biological tools for measuring beach erosion may be helpful for the interpretation of depositional processes of beach environment. The computation is on the basis of random data collection of modified burrow mouth, which was 3.5 cm in average diameter before being inundated and 6 cm diameter modified after inundation.

7.12.3 Computation

Area of the habitat of the red crab: $178\ \text{m} \times 41\ \text{m} = 7,298\,\text{m}^2$.

Volume of thin veneer of sands with organic matter: $178 \times 41 \times 0.09 = 656.82\,\text{m}^3$.

Volume of burrow mouth of almost filled up half sphere shaped burrow: $^2/_3\pi r^3$

(As modified burrow mouth is 6 cm in diameter, then radius is 3 cm)

$$= {}^{2}/_{3} \times {}^{22}/_{7} \times (0.03)^3$$
$$= 0.0000565 \ \mathrm{m}^3.$$

As an average of one burrow occurred in the entire area, then the number of such burrow in 7,298 m^2 would be 7,298 (approximately).

The volume of sediment loss in each inundation: $7,298 \times 0.0000565 = 0.412337 \ \mathrm{m}^3$.

The area with the crab habitat inundated twice daily for maximum 7 days in both full moon and new moon, the volume of sediment loss in a month: $(7 \times 2) \times 2 \times 0.412337 = 11.545 \ \mathrm{m}^3$.

(Here 7 days multiplied with 2 for inundation twice daily and another 2 for inundation in both full moon and new moon i.e. two times in a month)

The period for bio-erosion of thin sand sheet of crab habitat

$$= 656.82 \ \mathrm{m}^3 \div 11.545$$
$$= 56.89$$
$$= 57 \ \text{months} \ \ \text{(approximately)}$$

It is found that some part of the beach area is densely burrowed whereas burrows scattered in less number on the adjacent part of the beach depending upon the concentration of the organic matter as the crab, *Ocypode macrocera* is a benthic feeder looking for the organic particles from the beach sands. Thus, occurrence of crab burrows in and around the beach is random, irregular and not maintaining uniformity in their appearance. With the increasing wt of ejected materials number of burrows increased. Thus, it is positively correlated if is really difficult to relate the ejected materials of a particular burrow. Number of burrow denoted the population density of the crabs.

The formula is set for the calculation of beach materials due to bioturbation activities by the red crabs, *O. macrocera* (Fig. 7.7). This formula helps in the computation of the net sediment budget of the depositional processes particularly in the intertidal zones almost composed of beach sands. In this way the biological tool may be helpful on estimation of beach erosion.

7.13 Summary

Estuarine banks exhibit phenomenon of erosion as evidenced by slumping and terrace structure due to asymmetric tidal flows simultaneous with sediment compaction supporting vegetations in different stratigraphic levels. An asymmetric distribution of flow velocity over transverse profile of the estuaries facilitates erosion generally on the outer concave side and is the resultant from the combination of helicoidal and sinuous tidal flows. This flow complex suggests a lower partof the lower flow regime

Fig. 7.7 Grazing traces by red crab, *Ocypode macrocera*

to the upper part of the lower flow regime (Simon and Richardson 1962). It may be concluded that this process of estuarine bank erosion leads to changes of their geomorphology (Das 2010a). Islands formed by mainland cut off during bifurcation of channels in the estuarine region are the result of the shallowing of estuarine beds. This is further associated with intermittent subsidence and sedimentation in the estuaries. The active processes all along the coasts at the bottom of the estuaries are erosion and accretion. However, the large quantities of sediments that are arriving or departing the estuarine region to maintain the natural equilibrium still require investigations.

Erosive processes at tidal channels are mainly related to the flood current dominance. This leads to migration of the tidal channel features in one direction and dominant sediment transport is in the opposite direction as estuaries behave analogously to rivers (Dyer 1977). This process of river bank erosion and deposition leading to migration of their courses all over the deltaic plains of Sunderbans might have relations to the neo-tectonism of the Bengal basin (Das 2010a). Islands formed by mainland cut off during bifurcation of channels resulting from shallowing of river beds are associated with intermittent subsidence and sedimentation of the Bengal delta.

The Erosional features like furrow and ridge structures, mud balls and mud mounds help in understanding the morphodynamic processes that lead to the landward displacement of shoreline in the deltaic Sunderbans which reached its present position about 5,000 years BP.

Formations of islands are sometimes noticed as a result of estuarine erosion. Mid channel bar evolved by disconnection of the bank ward edge of the bar due to erosion is occupied by active or abandoned spill channels. Islands formed by mainland cut off during bifurcation of channels resulting from shallowing of estuarine beds. This is associated with intermittent subsidence and sedimentation.

References

Allen JRL (1984) Sedimentary structures, their character and physical bases. Development in sedimentology. Elsevier, Amsterdam

Bell HS (1940) Armoured mud balls-their origin, properties and role in sedimentation. J Geol 48:1–31

Bhattacharya A (1999) Embankments and their ecological impacts: a case study from the tropical low-lying coastal plains of the deltaic Sunderbans, India. In: Grann H, Vollmer M (eds) Large-scale constructions in coastal environments. Springer, Berlin, pp 171–180

Bhattacharya A (2000) Some geomorphic observations indicating shoreline variation in the coastal tract of West Bengal (India) – a case study around Bakkhali in Sunderbans. In: Proceedings of the international quaternary seminar on INQUA shoreline. Indian Ocean Sub-Commission, New Academic Publishers, Delhi, pp 32–37

Das GK (2008) Evidences of some erosional features along the river bank of Sunderbans, India. Geogr Rev India 70(2):139–145

Das GK (2009a) Beach processes of Tidal Islands of Hugli Estuary, West Bengal. Geogr Rev India 71(3):240–248

Das GK (2009b) Beach erosion for crab -burrowing. Indian Sci Cruiser 23(1):58–59

Das GK (2010a) Environmental status of Bakkhali Beach, West Bengal. Geogr Rev India 72(2):132–139

Das GK (2010b) Biodiversity in Sunderbans. In: Dwivedi AK, Tripathi SC (eds) Environment and biodiversity, environmental pollution and risk to biodiversity. Lambert Academic Publishing AG & Co., Saarbrucken, pp 1–34

Das GK, Bhattacharya A (1994) A piecemeal mechanism of bank erosion following subsidence: a case study from Thakauran river of deltaic Sundarban, West Bengal. J Indian Soc Coast Agric Res 12:231–234

Dyer KR (1977) Lateral circulation effects in estuaries. In: Officer CB (ed) Estuaries, geophysics of the environment-overview, summary and recommendations. Natl Acad Sci 22–29

Ginsberg SS, Perillo GME (2004) Characteristics of tidal channels in a mesotidal estuary of Argentina. J Coast Res 20(2):489–497

Gopinath G, Seralathan P (2005) Rapid erosion of the coast of Sagar Island, W.B. India. Environ Geol 48:1058–1067

Kilburn RN (1977) Description of new species of *Amalda Chilotygma* (Gastropoda: Olividae: Ancillinae) with a note on the systematics of *Amalda, Ancillus* and *Ancillista.* Ann Natal Mus 3(1):13–21

Marshall AJ, Williams WD (1982) Textbook of zoology, vol 1, 7th edn. Macmillan, London

Roychaudhuri S, Sengupta R, Rakshit S, Biswas NR (2003) Vulnerability of the West Bengal coastal zone. Newsl Mar Wing Geol Surv India XVII:33–35

Simon DB, Richardson EV (1962) Resistance to flow in alluvial channels. Trans Am Soc Civ Eng 127:927–953

Uehara K, Sojisuporn P, Saito Y, Jarupongsakul T (2010) Erosion and accretion processes in a muddy dissipative coast, the Chao Phraya River delta. Thail Earth Surf Process Landf 35:1701–1711

Chapter 8
Floral Diversity

Abstract Sunderbans forest encompasses a wide variety of halophytic mangrove species, salt marshes and sea shrubs. It constitutes a unique Biosphere Reserve in the coastal Bay of Bengal. The pattern of mangrove vegetation depends on the type of geomorphic set-up. Point bars and mid channel bars of the rivers of Sunderbans primarily support mangrove vegetation, whereas marsh vegetation is confined to upper intertidal to supratidal zones of levees and river flood plain. Species diversity of mangrove depend on the preferred salinity of respective species, availability of nutrients, organic carbon, soil texture, water retention capacity of soil etc. In general, *Excoecaria – Avicennia – Phoenix* combination is found to occur in the upper stretch of Sunderbans rivers, while *Avicennia – Aegialitis – Rhizophora* and *Avicennia – Ceriops – Rhizophora* combinations are seen in the middle and lowest stretch of the rivers respectively. In spite of this immense contribution as protector of shorelines as well as agricultural land, it is a matter of serious concern that local populace often do not realize the importance of mangrove forest and exploit it in an unplanned way. To save this unique Biosphere Reserve which exhibit various generic and species diversity, massive awareness programme is most essential.

Keywords Sunderbans • Mangroves • Species diversity • Salinity • Salt marsh • Afforestation • Deforestation

The estuarine delta complex of Sunderbans supports the world's most luxuriant mangrove forest, a well-known forest ecosystem of the tropics (Vannucci 1989). Mangroves are self-maintaining estuarine components of the biosphere that are in continuous interactions with terrestrial and marine ecosystems. The luxuriant and ever green mangroves vegetations cover the entire dynamic and changing environment of the coastlines, river flood plains, mouth of the river and delta fronts of the estuarine Sunderbans. Mangroves and salt marsh along the coastlines, islands, mudflats and on the gentle slopes of the tidal channels form the floral diversity of the deltaic Sunderbans. The mangrove swamps have a unique mosaic of habitats with an extremely rich diversity of flora and fauna and serve as a filter to the pollutants, protect coastal regions and supply nutrients to the adjacent ecosystems. Mangroves of the intertidal zone are inundated twice daily during high tides through a complex

G.K. Das, *Estuarine Morphodynamics of the Sunderbans*,
Coastal Research Library 11, DOI 10.1007/978-3-319-11343-2_8

network of tidal creeks. The vast saline to brackish watershed with a large number of siliciclastic islands is famous for its biodiversity and has been declared as a Biosphere Reserve by the Government of India in 1989.

The floral species of the Sunderbans mangroves swamps are not made up of one homogeneous species group; it has a unique heterogeneous combination of different species. The mangrove forests along the margin and marginal islands of the rivers stand as a productive and protective fence. This very vibrating ecosystem supports numerous terrestrial, benthic and aquatic organisms and displays complex association of species exchange and exchange of materials and energy within the system involving land and water. Activities of mangrove fauna in the intertidal flats bring about substantial topographical and morphological changes. Mangroves help trapping of sediments in the intertidal flats and promote stabilization of the substratum.

The dominant mangrove plant species are known for several adaptations to the halophytic environment including prop roots, salt exclusion, salt excretion, salt accumulation (Balsamo and Thomson 1995; Dschida et al. 1992; Azocar et al. 1992; Lin and Sternberg 1994) and production of viviparous seedlings. From the ecological point of view, these high-salinity tolerant plants respond variously to the change in the morphological and physiological conditions in different parts of the intertidal zones. Mangrove plants develop salt balance mechanism in their root levels, leaf levels as well as in their tissue levels. Further, their genetic potential too quickly adapts to changing environment caused by tidal and monsoonal cycles. Mangrove species usually have breathing roots i.e. the pneumatophores, stilt roots, crypto viviparus seeds and special mechanism of seed dispersals for their sustainability in such dynamic systems (Fig. 8.1).

Fig. 8.1 Adaptive features of mangroves: fruits of salt-tolerant mangrove species *Brugeira* sp. that undergo vegetative reproduction (*left*); pneumatophores (breathing roots) of *Xylocarpous* sp. (*right*)

8.1 Mangroves

Mangroves are woody plants that grow at the interface between land and sea in tropical and sub-tropical latitudes where they exist in condition of high salinity, extreme tides, strong winds, high temperatures and muddy, anaerobic soils. The word 'Mangroves' is of doubtful origin and there is no concrete record of when it has first come into use. It is usually considered to be derived from the Portuguese word 'Mangue' with the English 'grove'. In Portuguese, the word 'mangue' is utilised for mangrove trees, or bush; the mangrove community as a whole is called as 'mangal'.

8.1.1 Geological History of Mangroves

Mangroves existed even in the mega city, Calcutta in 9,000–8,000 year BP migrated southwards during 7,000–6,000 year BP (Umitsu 1997). The mangroves of Sunderbans grew over the deltaic plain which was formed not less than 10,000 year BP by the accretion of sediments carried by the Ganga – Brahmaputra – Meghna (GBM) river systems from the Himalayas. The surface geology of the upper portion reveals quaternary sediments intercalated with the sand silt and clay of marine origin. The Bengal basin, surrounded by the Indo-Barman ranges of Cenozoic era in the east, Pre-Cambrian Indian shield in the west and Shillong plateau in the north east characterises the common genesis of geological and tectonical relations of the deltaic Sunderbans (Allison 1998). Shifting of mangroves vegetations from the north to south is as a result of deforestation by the human habitation after the last glacial period of about 11,000 year BP where the temperature is considered to be about 6 °C (Tomlinson 1986). The mangroves flora and vegetations dynamics are considered as the indicators of the climatic condition of the Sunderbans. Thus, the geological history of the tidal tropical forest, Sunderbans is recognized with the help of bioclimatology that in or words mangroves remain as the focal indicators of climates of the then Sunderbans.

8.2 Forest Management

Sunderbans was first recorded as lands of 24 Parganas, the then jaigir of Lord Clive of the East India Company in 1757. After that the history of the Sunderbans was only about the change of land use pattern converting the forest areas into the agricultural lands and that was introduced by the then Collector General of 24 Parganas, Mr. Claude Russel in 1770. The Sunderbans forest has come under well planned and scientific management since 1970s of the last century. Earlier Heimig in the British era introduced the first working plan for the Sunderbans forest during the period 1893–1894 to 1902–1903 where permit was given strictly for cutting of sundari tree with the diameter more than 3 in. Lloyd, Farrington and Bart maintained this first working plan and executed this policy of felling trees in the couples annually up to 1903–1904 to 1911–1912. Trafford placed the second working plan for the period from 1912–1913 to 1931–1932 and the third one was made by Curtis for 1931–1932 to 1951–1952 but that 3rd working plan was not maintained from 1937 to 1938. Since then Mr S. Chowdhury worked out a plan assembling all circles into a single working circle from 1937–1938 to 1948–1949. This scheme was continued on 20 years felling cycles under selection cum thinning system. First working plan for managing forest of Sunderbans in the independent India was introduced in 1949–1950 for 10 years duration. No working plan was prepared after the end of this plan in 1958–1959 and the same plan was followed up to 1976–1977 for felling operation in annual coupe. Refuge problem at Marichjhapi created inconveniences for the implementation of this plans during the consecutive years from 1977 to 1980. The entire Sunderbans region is at present distributed between Sunderbans Tiger Reserve (STR) and 24 Parganas South Forest Division which are geographically separated by the Matla River. The Ministry of Environment and Forests, Government of India declared the entire 9,630 km^2 of Sunderbans as the Sunderbans Biosphere Reserve (SBR) in 1989 for conservation, management, research, training and maintaining harmony between man and environment of mangrove forest lands, Sunderbans (Anonymous 2003).

8.3 Mangrove Characteristics

Mangroves are self-maintaining estuarine components of the Sunderban Biosphere Reserve that are in continuous interaction with terrestrial and marine ecosystem. Salt-tolerant mangroves are physiological halophytes with specialized adaptations (Fig. 8.1) like vertical pneumatophores, buttresses, knee, cable, and halophilous properties with salt glands. The mangroves are also vivipary. The mangroves play a conspicuous role in this specialized environment by obstructing the direction of currents, accreting silt and binding the soil.

8.4 Geomorphic Set Up of Mangrove Swamps

Mangroves occupy three distinctive geomorphic zones:

(i) Elongated stretch of mangroves paralleling the discontinuous marginal levees: The transverse extension of these mangrove forests is often truncated by the embankment and revetments built for protection of cultivated land. The geomorphic character of such mangrove swamps is comparable to that described by Woodroffe (1999).
(ii) Lenticular patches of mangroves covering crescentic point bar deposits: These are commonly located at confluence of two rivers.

(iii) Sparse mangrove vegetation on the mud flats of mid-channel bar: Both overlapping and non-overlapping nature of mangroves are noticeable. Mangroves of different species in a forest generally overlap each other. However, *Sonneretia* sp. maintains a unique identity of its own by its appearance in the lowest inter-tidal zone and it exhibits a distinct non-overlapping character from the major cluster of other mangroves. With time there is a gradual filling up of the space by the appearance of other species of mangroves. *Sonneretia* sp. thus intermingles with other mangrove species to form a cluster (Das and Bhattacharya 1991c).

Species diversity occurs in different parts of the intertidal flats along the longitudinal profile of the rivers and their distributaries, eg. (I) *Excoecaria-Avicennia-Phoenix* combination –in the upper stretch of the river, (II) *Avicennia – Aegialitis-Rhizophora* combination in the middle portion and (III) *Avicennia-Ceriops-Rhizophora* combination in the lowest stretch of the stream.

8.5 Floral Dynamics

Floral dynamics are controlled by several factors like different site availability, different specie availability, water quality, climatic conditions and different species performances (Pickett and Mary 2005). In the Sunderbans, the mangrove vegetation dynamics have been changing principally due to reduced fresh water supply into the estuarine river water and other related factors.

Mangroves occur in the different locations of Sunderbans depending upon the salinity gradient. Chronic high salinity is detrimental to the mangroves although the mangrove plants have some tolerance power of salinity. Hyper-salinity stunts the growth of the tree, reduces biomass and causes denaturing of buds. Higher concentration of salt affects plant growth by accumulation of ions in toxic concentration with the plant tissue.

Sunderbans mangroves wetland has been recognized as highly saline zone due to dearth of fresh water supply. Salinity controls the survival, growth and succession, species composition, distribution of plant species in the Sunderbans. Topography, climate, hydrology, tidal fluctuation controls this salinity of estuarine river waters of Sunderbans. In lower saline areas the mangroves are seen more luxuriant as the growth of mangrove forests and productivity are affected by salinity.

8.6 Forest Types

The Sunderbans forest has been classified into different types from time to time by so many workers based on salinity tolerances, mangrove habitats and geomorphic situations. Systematic zonation of mangroves vegetation was first introduced by

Sir David Prain in 1903 (Agharkar and Ghosh 1931; Anonymous 1935). He categorized the mangroves of Sunderbans of undivided India into three types:

(i) South-west zone and southern coastal strip with well known dominant mangrove species
(ii) Central zone habituated by *Heritiera fomes* vegetations
(iii) North-eastern part of savannah type of vegetations with salt marshes and sea grasses

After 30 years of Prain's classification, Curtis divided the mangrove vegetations into three types:

(i) Fresh water vegetations
(ii) Salt water vegetations with moderate salinity
(iii) High salinity tolerant vegetations

In 1936, Griffith classified the mangrove forest of the Sunderbans (Anonymous 1954) into two main categories:

(i) Mangrove swamps
(ii) Coastal sand dune vegetations

Further in the same year of 1936, Champion arranged the forest of Sunderbans (Anonymous 1954) into four types:

(i) Fresh water forest
(ii) Salt water mixed forest
(iii) Mangrove forest of tree categories
(iv) Brackish water mixed low saline mangrove forests

Later Champion and Khattak (1965) classified the Sunderbans forest as moist tropical forest comprising a mosaic of beach forest and tidal forest. After a long and rigorous survey of the entire zone of the Sunderbans, the forest type may be classified into four categories concerning habitat of mangrove occurrences. It is more relevant to consider the following category of classification as the mangrove zonations and that will be more appropriate forest type of the Sunderbans.

(i) Namkhana – Sagar Island zone
(ii) Basanti – Sonakhali zone
(iii) Hasnabad – Sandeshkhali zone
(iv) Gosaba – SBR zone

All these categories again sub divided into several types based on the vegetations dynamics:

(i) Coastal vegetations including sand dune creepers
(ii) Point bar and mid channel bar vegetations
(iii) River flood plain and mud flat vegetation including salt marsh and shrubs
(iv) Three-tier tidal creek vegetations

Heritiera fomes, a fresh water species is now almost rare in the Sunderbans of the Indian part. *Excoecaria agallocha*, *Ceriops decandra*, *Xylocarpus* sp., *Sonneratia apetala* etc. are the moderate saline water species with wide ecological amplitude. Avicennia sp., Rhizophora sp., Agialitis rotundifolia etc. are saline water species with narrow ecological amplitude. Thus, mangroves may be classified into three categories based on the degree of tolerance of salinity:

(i) High salinity tolerant species
(ii) Medium salinity tolerant species
(iii) Low salinity brackish water species

8.7 Mangrove Varieties

In most of the areas of Sunderbans, mangrove forests are reclaimed for settlement due to human pressure. Large areas are brought into cultivable lands by felling mangrove trees and by protecting the hinterlands from salt-water menace by artificial daming.

The principal timber mangrove trees of Sundarbans are *Ceriops tagal*, *C. decandra. Excoecaria agallocha*, *Avicennia officinalis*, *A. alba*, *A. marina*, *Bruguiera gymnorrhiza*, *Sonneratia apetala*, *S. griffithii*, *Rhizophora mucronata*, *Aegialitis rotundifolia*, *Aegiceras corniculatum* and a palm species *Phoenix paludosa* whereas among the minor vegetation *Sarcolobus globosus*, *Derris trifoliata and Acanthus ilicifolius* are the most important (Das and Bhattacharya 1991c). These mangrove species generally occupy the intertidal zones of newly developed point bars; mid-channel bars, river banks and flood plan areas. The sundari tree (*Heritiera fomes*) and golpata (*Nypa fruticans*) are very rare at present in the Indian Sunderbans (Fig. 8.2).

Fig. 8.2 Tidal mud flat vegetations of *Nypa fruticans*

The intertidal flats of the River basin display a sequential growth of mangroves, which can be, classified into stage I, stage II and stage III mangroves (Das and Bhattacharya 1991c, 1994). Plants which initially occupy the river flood plains and natural levees constitute stage I mangroves (i.e. mangroves of first generation). Subsidence of banks with simultaneous sediment compaction leads to the burial of stage I mangroves. Upright trunks of stage I mangroves are often seen in river sections. The subsequently developed levees and flood plains support stage II or the mangroves of second generation. Point bars that accrete thereafter and the mid channel bars at points of intermingling of ebb and flood currents supports stage III or the newly introduced mangrove saplings (Fig. 8.3).

Sub-horizontal wood seams of thickness within 30 cm are found at several places at a depth of about 3 m from surface on the river banks. These wood seams, with some interruptions, continue laterally for about 1 km or more. These partially decomposed wood seams are derived from the biochemical and geochemical changes of stage I mangroves with the passage of time and in depth of burial (Das 2008).

The mangroves have long since been called the "green wall of the sea" as they provide protection from the waves. Kolkata was saved from cyclones due to the fact that mangrove surrounded the Sunderbans. The existence of this dense mangrove forest is an assurance of safety for the urban metropolis in case of any cyclonic or storm surge.

8.8 Mangrove Zonation

In terms of their sequential growth mangroves in the river basin of the Sunderbans occur as (i) overlapping and (ii) non-overlapping types. *Avicennia* sp. *Excoecaria* sp., *Aegialitis* sp. *Rhizophora* sp. and *Ceriops* sp. are the dominant plants of the intertidal flat. These plants generally occur on gently sloping (2–3°) flood plains of clayey silt composition. In the forest areas, mangroves of different species generally overlap in occurrence (Cardinale 2011a, b). *Sonneratia* sp. however, keeps its separate identity in the lowest intertidal zone and exhibits a distinct non-overlapping character from the major cluster of other mangrove plants. With time there is a gradual filling up of the space by the appearance of other species of mangroves. In its subsequent stages of development *Sonneratia* sp. intermingles with other mangrove species to form a cluster of mangrove species (Das and Bhattacharya 1991b, c).

The vegetation types of the mangrove swamp of the Sunderbans have a unique heterogeneity combination of different species. The vegetation succession at present very significant as some species is gradually diminishing at slow rate. *Heritiera fomes* disappears from the mangrove land of the Indian Sunderbans. Species diversity occurs in different portions of the interdial flats along the longitudinal profile of the rivers of Sunderbans. These are (i) *Excoecaria-Avicennia-phoenix* combination in the upper stretch, (ii) *Avicennia-Aegialitis-Rhizophora* combination in the Middle stretch and (iii) *Avicennia-Ceriops-Rhizophora* combination in the lower stretch.

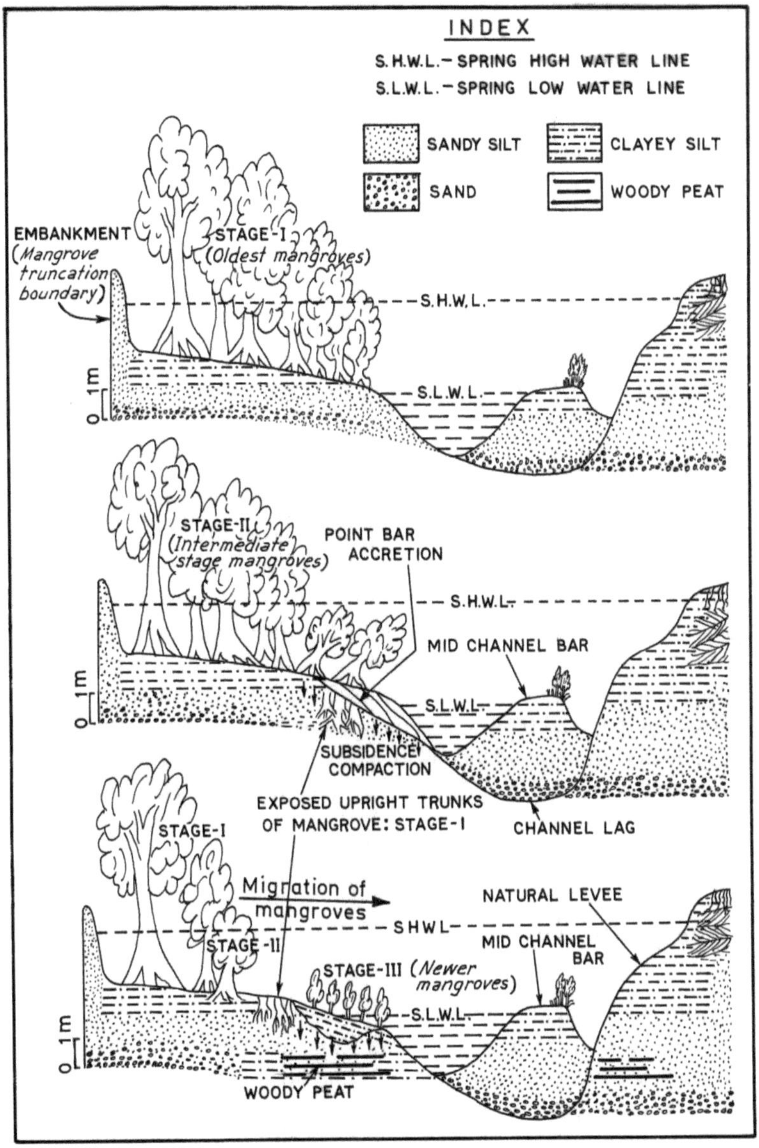

Fig. 8.3 Model showing the sequential appearance of mangroves on the river bank of Sunderbans

8.9 Habitat of Occurrence

Intertidal zones of point bars and mid-channel bars in the rivers of Sunderbans primarily support mangrove forests, whereas, marsh vegetation is confined to the upper intertidal to supratidal zones of natural levees and river flood plains. Of these, the accreting point bars are the most suitable sites for the mangrove vegetation. Margin of mid-channel bars comes next in preference. Subaqueous to subaerial levees and distal flood plains are the favourable sites for the formation of tidal marshes in the Sunderbans.

8.10 Chemistry of Mangrove Sediments

The study area typically belongs to the tropics with average maximum temperature of winter and summer months ranging from 20 °C to 30 °C, respectively the annual rainfall between 1,900 and 2,100 mm. The sediment salinity of the study area ranges between 4.93 and 5.8 ppt and the pH is slightly alkaline ranging between 8.0 and 8.3. Chemical analysis of sediment samples reveals that the sediments are enriched with nutrients. Estimation of NPK values revealed that available nitrogen (Av. N) ranges from 880 to 1,540 kg ha^{-1}, available phosphorus (Av. P) from 11.4 to 26.0 kg ha^{-1} and available potash (Av. K) from 2,083 to 2,957 kg ha^{-1}. The organic carbon shows a range between 0.38 and 0.69 % and the total organic matter in sediments 0.066–1.19 % (Table 4.3). It is found that available potash is five times richer than that of the required amount for the natural growth of mangroves, which supply the nutrients to the ambient environment.

8.11 Composition of Mangrove Sediments

Texturally, the sediments of the mangrove substratum are composed of 90 % of silt with subequal proportion of fine sand and clay constituting the rest 10 % of the bulk. The graphic mean size (Mz) of the sediment belongs to silt fraction (6.74 phi) and the sediments are well to moderately sorted ($\sigma_1 = 0.31$–1.4 phi). Most of the samples show a slight tendency of negative skewness with kurtosis (K_G) values often greater than 1.0 (Table 6.5). In general, the mangrove substratum sediments are restricted to a short range of grain size-sorting values and reflect a high sensitiveness of their depositional environment (Das and Bhattacharya 1991c; Bhattacharya 1999). The slight negative skewness (SK_1) of sediments indicates an admixture of coarse fraction derived from the mangrove plants to the major silt population in grain-size distributions. The good sorting and high peakedness of the distributions reflected their very limited size variation and dominance of silt population with respect to the sand and clay fractions respectively.

8.12 Salt Marshes

Marsh is a frequently or continually inundated wetland characterised by herbaceous vegetation (Fig. 8.4). Salt Marsh is a relatively soft, wet flat area that is periodically flooded by saline water where fine sediment is deposited and salt tolerant grasses grow. Marsh vegetation occurs sporadically all along the river margin, particularly in the muddy upper intertidal zone to the supratidal zone. They generally occupy areas above mangrove swamps and are typical of intertidal zone behind embankments or earthen dams constructed for protecting inland areas from salt-water menace. In addition to their diverse ecological and economical importance marsh vegetation has a special role as sediment baffles (Davis 1983; Das and Bhattacharya 1991c).

Marsh vegetation of the Sunderbans is characterised by creepers like *Suaeda nudiflora*, *S. maritima*, *Sesuvium portulacastrum* in low-lying areas of upper intertidal flats. The marsh grass, *Porteresia coarctata* occurs in patches and appears as a pioneering species in the newly deposited mudflats (Fig. 8.4). The beds of *Porteresia* sp. keeps a sharp contrast with that of *Suaeda* sp. in a transverse profile in the upper marshy zone with a compact mudy substratum, *Suaeda maritima* forms a stable matting even on the dessicated ground whereas, in areas of loose unconsolidated sediments near embankments *Suaeda nudiflora* appears with more heights and greater density. Population of *Porteresia coarctata* having heights upto 30–60 cm marks the lower marshy zone.

An altogether different flora is encountered in the dry supratidal areas on both banks of Sunderban Rivers. The vegetation includes grasses like *Paspalum* sp., *Panicum* sp. and *Aeluropus lagopoides*. These species further upward are followed by *Sesuvium portulacastrum*, *Ipomoea pes-caprae*, *Salicornia brachiata*, *Scirpus* sp. and *Hygrophyllux* sp. (Das and Bhattacharaya 1991a).

Fig. 8.4 Salt marsh vegetations (*Porteresia coarctata*) in the river flood plain

8.12.1 Texture and Chemistry of Marsh Sediments

Texturally, the marsh sediments are composed of more than 90 % of silt with admixtures of small proportions of fine sand and clay (Table 6.4). Graphic mean size (M_Z) belongs to silt fraction with moderate to poor sorting of grains. Most samples show a tendency of slightly negative skewness, which is due to admixture of coarser vegetable debris to the silt fraction. Sediment substrates of mangrove swamps and marsh do not show much variation and match well to that of the lagoon or distal shelf (cf. Friedman and Sanders 1978).

The marsh sediments display a salinity range between 3.44 and 5.36 ppt during the pre-monsoon time; the salinity drops down to 2.86–3.44 ppt during monsoon period. pH during these three seasons varies between 6.73 and 7.69 and does not show appreciable variations seasonally (Table 4.2). Organic carbon and Total organic matter in the sediments are 0.04–1.2 % and 0.07–2.06 % respectively and are close to that of the values of mangrove substratum sediments (Table 4.2).

8.12.2 Marsh as Sediment Binder

The vegetation carpet in the marshes and inhabiting fauna therein add to a special ecological attribute. Macrobenthic animals like gastropods, bivalves and crabs cause biological recycling of marsh sediments. The swamp and marsh vegetation imparts a baffling effect on the flood plains of River basin against the waves and currents. The marshes suffer inundation during high water spring tides when suspended particles of very fine sands, silt and clay are trapped by the swamp and marsh vegetation and settle down gradually. The efficient baffling and sediment trapping mechanism by marsh vegetation impart a greater coherence for the settled down particles (Das and Bhattacharya 1991a). As a result, during exceptionally high meteoric waves and cyclonic storms the muddy sediments of the marshes can escape erosion.

8.12.3 Roles of Salt Marsh

Salt marshes play several important roles in the mangrove ecosystems:

(i) They trap sediments and help enhancing accretion.
(ii) Salt marshes help capture of chemical elements including trace metals (Lacerda 1998; Costa and Davy 1992).
(iii) They enrich food resources by producing epizoonts for the fin fishes and shell fishes as foods.
(iv) They serve the role as nursery ground for a number of fish species.
(v) Salt marshes attract both shell fish and fin fish for abundant nutrient availability and enrich biomass.

8.13 Mangrove Afforestation and Deforestation

Mangroves play an important role in protecting the shorelines as well as agricultural lands as living green wall from major winds or waves induced erosion even from the tsunamis which recently struck (December 26, 2004) the coastline of Andhra Pradesh and Tamil Nadu. Increase of population in the mainland has resulted in the deforestation of Sunderbans mangrove for land reclamation human settlement. Afforestation programme for mangrove generation as well as the conservation has been undertaken in order to save Sunderbans. The Sunderbans Development Board introduced mangrove afforestation through artificial regeneration with the financial assistance of International Fund for Agricultural Development (IFAD) since 1982. These plantations covered the crescentic point bars, intertidal riverbanks, natural levees of tidal courses and inlets in the Sunderbans.

8.13.1 Regeneration

The experimental study of Sunderbans Development Board reveal that the artificial regeneration of mangrove in the mudflats are possible and the Board's effort on mangrove seed sowing through helicopters on the river banks in and around Sunderbans in 1991 was effective and its further extension appears to be prospective both from technical and economical consideration. The principal species for mangrove afforestation are *Avicennia* sp., *Bruguiera* sp., *Ceriops* sp., *Sonneratia* sp., *Rhizophora* sp. This artificial regeneration of mangroves in order to save this valuable genepool in perpetuity will be successful through active cooperation and participation of local inhabitants and panchayet raj system. Only panchayet will check the premature exploitation of mangroves.

Mangrove regeneration can be classified into two types – (i) natural and (ii) artificial regeneration. As natural regeneration of mangrove vegetation is limited, artificial regeneration i.e. plantation is taken into consideration (Redondo-Go'mez et al. 2008; Espinar et al. 2004, 2005) for several reasons – (a) cleared large areas are important for many observations after reforestation (b) lack of proper knowledge to improve growth of mangrove species and (c) information on artificial plantation is very limited.

8.13.2 Afforestation

Mangrove species growth in the natural conditions depends upon some environmental parameters mainly on soil/water salinity and tidal fluctuations. It is important to examine the soil profile, tidal fluctuations, water salinity etc. of the particular area where the mangrove plantation is being planned.

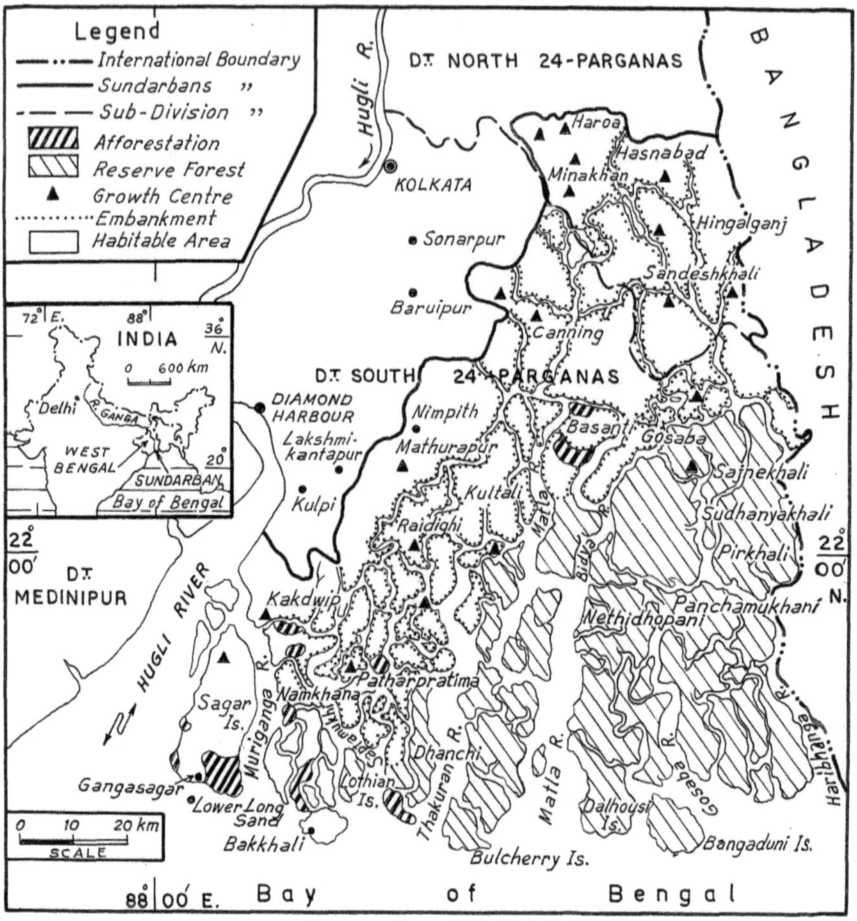

Fig. 8.5 Map of Sunderbans showing afforestation and reserved forest areas

Afforestation of mangroves in the sanctuary of Sunderbans (Fig. 8.5) through direct plantation of mangrove propagation or seeds was introduced first. Of late attempts are made to develop intertidal mangrove nursery and this attempt is becoming successful as because number of requisite saplings could be germinated depending upon the size of afforestation area.

A number of criteria required for development of mangrove nursery are as follows:

(i) An upper intertidal mudflat with gentle sloping.
(ii) The areas must be protected from strong tidal current, waves or surges.
(iii) Mudflat should be composed of clayey silt sediments.
(iv) The nursery site will be inundated on regular basis.

Required number of polybags of size 20 × 10 cm filled with mud collected from intertidal mudflat kept in upright serially at the mangrove nursery close to each other. Fully matured seeds, fruits, propagules of different species are collected from the mangrove forest. The collected materials will be sorted out in order to avoid infected, tender or injured seeds, fruits or propagules. This precaution results the decreasing mortality rate as well as the healthy growth of the saplings as mangrove trees in future. The mature seeds, fruits, propagates are placed gently in the mud of the polybags with tapering ends. It is advised to use separate beds for different mangrove species. Seeds and propagules placed in the polybags are found to be 98–100 % germinated. The dormancy period are varied for different mangrove species. For eg seeds of *Avicennia* grows rapidly whereas *Ceriops*, *Bruguiera*, *Sonneratia*, *Heritiera*, *Xylocarpus* take longtime for germination. Seeds propagules of some mangrove species like *Rhizophora*, *Bruguiera*, *Kandelia*, *Ceriops*, *Avicennia*, *Sonneratia*, *Aegiceros*, *Acanthus*, *Heritiera* were placed in polybags filled with non-saline soil and irrigated with fresh water. It is observed that the germination is quick and saplings are healthy in the polybags of freshwater irrigated non-saline agricultural land soil in comparison to those of the polybags filled with mud from mangrove substrates and irrigated with the brackish water. It may be said that the germination in initial stage is much better in fresh water condition than that of brackish water situations.

The 15–36 cm grown mangrove saplings are transplanted at low tide during the rainy season generally as because the soil and water salinity is comparatively low at the monsoonal time which causes higher survival rate although the seedlings can be transplanted throughout the year. Mangrove afforestation will be cent percent successful if the mangrove saplings are transplanted in the mesohaline and oligohaline regions with perennial freshwater supply.

The entire transplantation process is to be accomplished within a short span of time during low tide at ebbtime. Before transplantation some criteria should be observed such as site selection, tidal inundation level and mud substratum. Measuring all these, the plantation pattern are worked out and accordingly the saplings from the nursery are carried to the plantation site. The distance between the two rows and also two seedlings is maintained at 1–1.5 m and after completing about 15 rows, a gap of 10 m was left for future forestry operation or fishing, crab capture etc.

Mangrove afforestation programme is mainly labour oriented. So, therefore, the total cost varies from place to place. Mangrove plantation is raised in intertidal zones and mudflats. An area of more than 8,000 ha has been covered by mangrove plantation under different schemes.

8.13.3 Deforestation

Mangroves are salt tolerant forests flourishing on sheltered tropical coastlines. Mangrove forests are variously described as 'coastal woodlands', tidal forests' as

Fig. 8.6 Mangroves deforestation and transport of wood logs at Sunderbans

this vegetation grows at the meeting points of land and estuary. But the beautiful mangrove forest of Sunderbans shows a steady decline in mangrove species due to several factors as follows:

(i) Local people continue to fell mangroves plants (Fig. 8.6) as before for fuel wood, fodder, small timber and commercial wood.
(ii) People capture prawn broods through netting mostly during ebb tide and fish in the creek and insets and thereby destroy the mangrove saplings in the mudflats.
(iii) Sunderbans Project Tiger has taken more importance than mangrove afforestation naturally or artificially.
(iv) Crocodiles and Turtles Park in the sanctuaries are commercially more important than the ecological concern of mangrove preservations.
(v) Indiscriminate human settlement, legal and illegal land occupation for cultivation and prawn culture farm.
(vi) Illegal settlement of the immigrant people intruding from the neighbouring countries.
(vii) Populist development activities in and around mangrove areas by the political parties results in fast disappearances of mangrove forest.
(viii) Newly planted saplings of mangrove afforestation programme are frequently uprooted during prawn seed collection and capturing operations, navigations, grazing of bovides etc.

8.13.3.1 Measures for Deforestation

Legal administrative measures may be taken to check the fast mangrove deforestation as following:

(i) Provide the local people with newer non-forest avenues for employment and income generation so as to avoid their dependence on mangroves.
(ii) Needs for the fuel wood and small timber can be met by opening of wood depots by forest departments within the reach of the villagers.
(iii) The fishermen may be provided with better fishing nets and boats and be allowed to fish in the rivers outside the mangrove areas.

8.14 Summary

Most of the mangrove species prefer low saline substratum. The mangrove species inundated twice daily by tides prefer hyper-salinity soil areas whereas species those occur in the centre portion of islands prefer their habitat with low salinity. Mangrove species diversity and richness depend upon the preferred salinity of respective mangrove species. Mangrove zonation is happened in response to the salinity of both soil and water. Species diversity of mangroves is controlled not only by the salinity alone but by the several factors like availability of nutrients (N P K), organic carbon, soil texture, water holding capacity of soil etc.

Major populace dwelling in the Sunderbans are engaged in deforestation in the course of farming and fishing to enhance the domain. Riverbed sedimentation and accretion causing inundation and overflow of the creeks, rivers, and estuary is an important reason of decline of the world famous Sunderbans mangroves. Survey shows that during the last century more than 5,000 km^2 of mangrove forest area were cleared to make way for human settlement. Several hectares of mangrove zones at Basanti, Gosaba, Sahebkhali, Sandeskhali, Canning, Jharkhali, Patharpratima, Kakdwip, Namkhana, Raidighi, Rakhaskhali, Buraburir tat, Bakkhali, Sagar Island, Maipita, Kaikhali, Jambudweep and other areas are on the verge of decline. Both the state and the central government should be more active to protect mangrove forests from being destroyed. Apart from implementation of law, awareness campaign to restore ecosystem should be strengthened. The state government at present has launched awareness campaign highlighting the current role of mangroves to provide safety against the natural disasters like tsunamis and cyclones.

References

Agharkar SP, Ghosh AK (1931) The composition of Bengal flora. In: Proceedings of the Indian Science Congress, Nagpur

Allison MA (1998) Geological framework and environmental status of the Ganges- Brahmaputra delta. J Coast Res 14(3):826–836

Anonymous (1935) The forest of Bengal. Calcutta
Anonymous (1954) Forestry in Bengal. Calcutta
Anonymous (2003) A handbook on Sunderbans Biosphere Reserve. Ed. Director, SBR, Calcutta
Azocar A, Rada F, Orozco (1992) Relaciones hidricas e intercambio de gases en dos species de mangle, con mecanismos contrastantes de regulacion de la salinidad interna. Ectropicos 5(2):11–19
Balsamo RA, Thomson WW (1995) Salt effects on membranes of the hypodermis and mesophyll cells of Avicennia germinans: a freeze fracture study. Am J Bot 82(4):435–441
Bhattacharya A (1999) Embankments and their ecological impacts: a case study from the tropical low-lying coastal plains of the deltaic Sunderbans, India. In: Vollmer M, Grann H (eds) Large-scale constructions in coastal environments, Springer, Berlin, pp 171–180
Cardinale BJ (2011a) Biodiversity improves water quality through niche partitioning. Nature 472:86–89
Cardinale BJ (2011b) The functional role of producer diversity in ecosystems. Am J Bot 98:572–592
Champion HG, Khattak GM (1965) Manual of silviculture for Pakistan. Pakistan Forest Institute, Peshawar
Costa CSB, Davy AJ (1992) Coastal salt marsh communities of Latin America. In: Seeliger U (ed) Coastal plant communities of Latin America. Academic, San Diego, pp 179–199
Das GK (2008) Evidences of some erosional features along the river bank of Sunderbans, India. Geogr Rev India 70(2):139–145
Das GK, Bhattacharya A (1991a) Sediments baffles by inter tidal Marsh and swamp vegetation: a study from the distributary channel banks of the Thakuran Tidal River, West Bengal. J Indian Soc Coast Agric Res 9:399–403
Das GK, Bhattacharya A (1991b) Pattern of sedimentation with reference to erosional and depositional areas along meander bends in the upper estuarine stretch of the Thakuran River, West Bengal. J Indian Soc Coast Agric Res 9:37–42
Das GK, Bhattacharya A (1991c) Physiographic set up and sensitiveness of textural parameters of tidal river related mangrove sediments. In: Proceedings of the National symposium on conservation and management of Mangroves. Bidhan Chandra Krishi Viswavidyalaya, Kalyani (WB), pp 46–52
Das GK, Bhattacharya A (1994) A piecemeal mechanism of bank erosion following subsidence: a case study from Thakauran river of deltaic Sundarban, West Bengal. J Indian Soc Coast Agric Res 12:231–234
Davis RAD (1983) Depositional systems. Prentice-Hall, INC., Englewood Cliffs
Dschida W, Platt-Aloia K, Thomson W (1992) Epidermal peels of *Avicennia germinans* (L.) Stern: a useful system to study the function of salt glands. Ann Bot 70(6):501–509
Espinar JL, Garcia LV, Figuerola J, Green AJ, Clemente L (2004) Halophyte germination in Mediterranean wetlands: gut-passage by ducks changes seed response to salinity. J Veg Sci 15:313–320
Espinar JL, Garcia LV, Clemente L (2005) Seed storage conditions change the germination pattern of clonal growth plants in Mediterranean salt marshes. Am J Bot 92:1094–1101
Friedman GM, Sanders JE (1978) Principles of sedimentology. Wiley, New York, 792p
Lacerda LD (1998) Trace metals biogeochemistry and diffuse pollution in mangrove ecosystems. ISME Mangrove Ecosyst Occas Pap 2:1–61
Lin GH, Sternberg LDSL (1994) Utilization of surface water by mangroves (Rhizophora mangle L.): an isotopic study. Bull Mar Sci 54(1):94–102
Pickett STA, Mary LC (2005) Vegetation dynamics. In: Eddy (ed) Vegetation ecology. Blackwell Publishing, Malden, pp 172–195
Redondo-Go'mez S, Mateos Naranjo E, Garzo'n O, Castillo JM, Luque T, Figueroa ME (2008) Effects of salinity on germination and seedling establishment of endangered *Limonium emarginatum* (Willd.) O. Kuntze. J Coast Res 24(1A):201–205
Tomlinson BP (1986) The botany of mangroves. Cambridge University Press, London

Umitsu M (1997) Landforms and floods in the Ganges delta and coastal lowland of Bangladesh. Mar Geol 20(1):76–87
Vannucci M (1989) The mangroves and us; a synthesis of insights. Indian Association for the Advancement of Science, New Delhi, 203p
Woodroffe CD (1999) Response of mangrove shorelines to sea level change. Tropics 8(3):159–177

Chapter 9
Mangrove Ecosystems

Abstract A combination of estuarine and mangrove forest ecosystem is the characteristics of the Sunderbans region. Natural processes like tidal surge, cyclones etc., acting singularly or in combination, have made the Sunderbans mangroves ecosystem most fragile and harsh. This large block of halophytic mangrove ecosystem is highly productive which provides large quantities of organic detritus which is considered as primary food for the natural habitats of this ecosystem. The close relationship between microbe – nutrient – plant acts as a mechanism to recycle and conserve nutrients in the mangrove ecosystem. Hydrological, biogeochemical and ecological functions of mangrove have direct bearing on socio-economic benefits of the region. The mangrove ecosystem of Sunderbans is highly disturbed due to over exploitation of mangrove forests and unplanned land use pattern.

Keywords Sunderbans • Ecosystem • Mangrove • Food web • Detritus • Microbe • Nutrients

Mangroves are tropical phenomenon. These mangroves vegetations of typical halophytic adaptive features is found only in the tropical and subtropical coasts around the world and cannot grow in the cold region below 20 °C temperature. Estuarine, riverine and tide-influenced mangrove ecosystem of Sunderbans is dynamic and open in nature. The mangrove ecosystem is fragile and the ecological balance is unstable. Its geographical location as well as topography is unlike from that of other ecosystems in the world. Here the biotope is very much unique. Estuarine river waters are saturated with the salinity that helps occurrences of the mangroves species. The deltaic Sunderbans stands within the 30° N of equator where the average solar radiation is above 500 cal daily that helps decomposition of mangrove litters well. The potential of more than 500 KWH of the solar radiation hastens transformations of organic matter in soil. The deltaic soil has the special textural and structural characteristics that assist in the process of fast desalination in the water logging situation through precipitation. The tidal rivers and creeks are alternately inundated twice daily and only for a short duration the creek beds are exposed. This natural phenomenon helps on the abundant occurrences of micro and macro flora and fauna in the Sunderbans that could only thrive under this unusual changing morpho-dynamic situation. Mode of transport of matter makes this ecosystem as

G.K. Das, *Estuarine Morphodynamics of the Sunderbans*,
Coastal Research Library 11, DOI 10.1007/978-3-319-11343-2_9

open system through the flow characteristics and pathways type. Rainfall, tide, terrestrial run off litter production, decomposition etc. control these pathways as well as the rate of transport of nutrients in both ways and storage of materials. These flows and transformations import the organic matter from terrestrial ecosystems and export the particulate and dissolved organic matter to the adjacent marine ecosystems.

Mangrove ecosystem has the significance as the coastal stabilizers, shelterbelt areas and nursery grounds for aquaculture, production of tannin, timber, paper and pulp, charcoal and several other by-products. Mangroves enhance land accretion, stabilize embankment and act as the green wall of the sea as they provide protection from the extensive damages caused by the waves and tsunamis. Mangroves forest known other way as 'Tidal forest' or 'Coastal wetland' has the ecological, economic and social significance. Mangrove ecosystems maintain the most important links between the land and the sea particularly along the deltaic estuaries.

9.1 Mangrove Community

Mangrove community as a whole is termed as '*Mangals*'. Occurring in the intertidal mudflat, the mangroves is an assemblage of plants of different species with special halophytic adaptive features such as pneumatophores, stilt roots, root buttress, salt gland, viviparous germination etc.

Self-sufficient detritus based mangrove ecosystem produces and utilizes the food materials within its periphery. The complicated network of mangrove roots baffles sediments. These roots reduce the wave action. Thus the function of mangroves is to protect the agricultural land by minimizing the coastal erosion, reducing the atmospheric depression, tsunamis, storms, and cyclones, tidal bores and surges.

The natural habitat for growth and reproduction of shell fish, finfish, mollusks, and crabs of great economic importance is the detritus based nutrient rich mangrove ecosystem. Mangroves of typical halophytic adaptive features are found only in the tropical and sub tropical coasts around the world and cannot grow in the cold region below 20 °C temperature as mangroves are tropical phenomenon.

9.2 Mangrove Fauna and Food Web

Mangroves are the home for endangered wild species like estuarine crocodiles, otters, fishing cats, dolphins, pythons and cobra and other species. The mangroves fauna are unique and depend upon the mangrove habitats, even the crowns of trees including trunk, branches, leaves, flowers and fruits that provide a niche essentially to terrestrial fauna. The conceptual model of the mangrove ecosystem diagram shows the interactions and the interconnection among fresh water inflow, nutrient uptakes, tidal flows, consumers, physic-chemical parameters and the animal sediment relationship.

The food web of the Sunderbans mangrove ecosystems is the characteristic of shallow muddy estuaries with an extensive plant community of sea grasses, salt marshes, mangroves and macro algae. This type of mangrove food web is framed by relatively small number of path ways along which energy flows. The input of energy is principally in the form of mangrove detritus but always with a component of fresh benthic micro algae. The key organisms in this food web are a group of detritus consumers, herbivores and omnivorous crustaceans, mollusks, insect larvae, nematodes, polychaetes and a few fishes. The seasonal variation in salinity has the effects over the mammals, reptiles, amphibians to fishes and crabs, prawn etc. Birds of about 120 species are attracted as the mangroves ecosystem provides food in the forms of fish, crabs, mud skippers, insects especially in the ebb tidal situation when the river flood plain is exposed for the animal movement after the recession of water.

9.3 Primary Food Source

Detritus, derived from the decomposed mangrove litters are considered as the primary food source in the mangrove ecosystem. Detritus is the materials resulting from the disintegration of dead organic remains. Rate of decomposition of mangrove litters is very slow. Fungal and bacterial attack on mangrove leaves and the sequential change thereon of decomposition of leaves are thoroughly observed (Table 9.1).

The natural habitat for growth and reproduction of shellfish, finfish, mollusks, crabs of great economic importance is the detritus based nutrient rich mangrove ecosystem. Detritus is found as the major component inside the stomach content in both the fin fishes and shell fishes (Figs. 9.1 and 9.2).

Table 9.1 Components of different trophic levels in the mangrove ecosystem of Sunderbans

Producers:	Phytoplankton, algae, sea grasses salt marsh, mangrove
Herbivores:	Ciliates, meiofauna, insect larvae, molluscs, mullets, copepods
Omnivores:	Meiofauna, polychaetes, shrimps amphipods, molluscs, fish
Primary carnivores:	Fin fish larvae, small finfish, eels, catfishers, soles
Secondary carnivores:	Catfishes, birds (herons, waders, sea gulls)
Tertiary carnivores:	Finfish (perches, tarpons), birds (eagles), sharks
Decomposers:	Fungi, bacteria, protozoa etc.

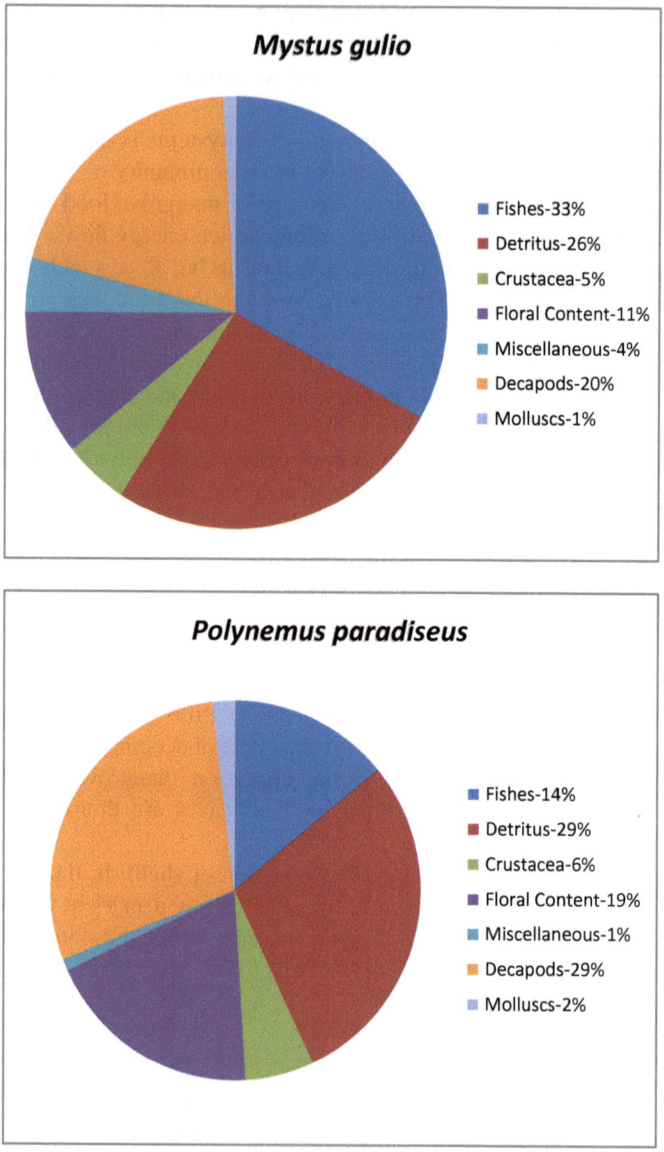

Fig. 9.1 Detritus part in the stomach content of mangrove fin fishes

9.4 Components of the Mangrove Ecosystems

Producers are the green plants, which are able to synthesize complex organic compound from inorganic substances by photosynthesis in presence of sunlight, phytoplankton, micro and macro algae, sea grasses, salt marsh, and mangroves are

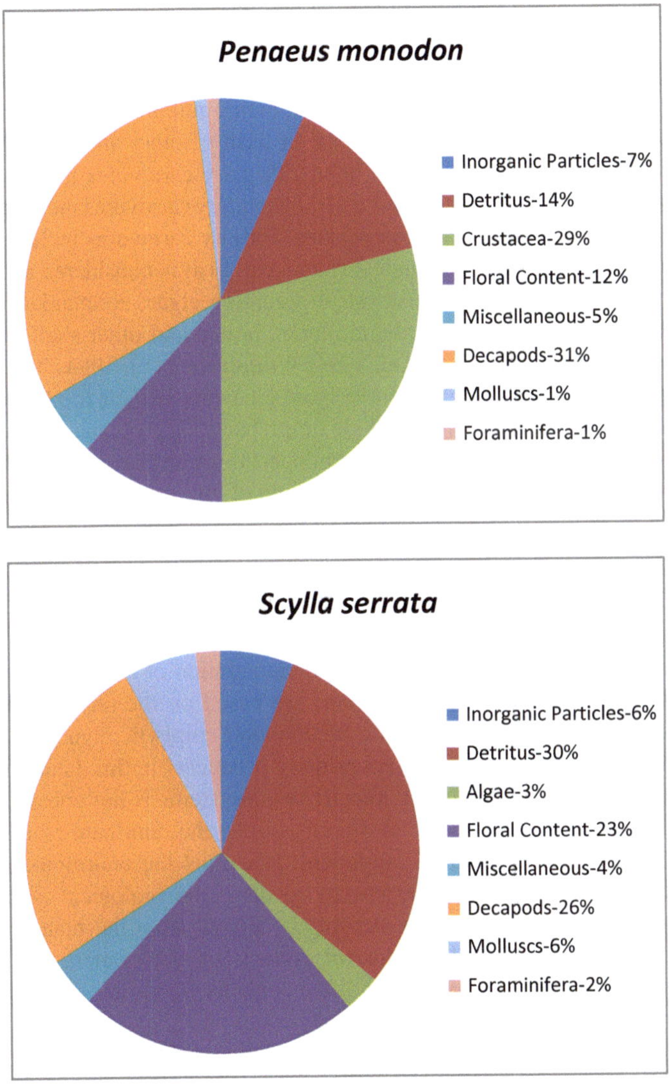

Fig. 9.2 Detritus percentage in the stomach content of mangrove shell fishes

the producers in mangrove ecosystem. Algae play an important role as producer in the mangrove ecosystem. The horizontal and vertical distribution pattern of mangrove algae depend upon some environmental factors like tidal fluctuation, wave action, salinity, temperature, light penetration and intensity, wetting frequency, desiccation, turbidity etc. (Phillips et al. 1994; Farnsworth and Ellison 1996). Consumers, the animals which take organic matter or other animals as food on the basis of organisms consumed or eaten, are of three types in the mangrove ecosystem as –(i) Herbivores (ii) Omnivores and (iii) Carnivores. Herbivores are the animals, which entirely depend on plants for their food and energy. They include ciliates, Meiofauna, insect larvae, molluscs, mullets, copepods etc. in this ecosystem. Omnivores are the animals that feed on both plants and animals. Omnivores include polychaetes, shrimps, meiofauna, amphipods, molluscs, fish etc. Carnivorous animals are those which feed on flesh of other animals. They are of three types on the basis of their food habits. Primary carnivores include small fin fish, fin fish larvae, eels, catfishes, and soles. Secondary carnivores include birds like herons, waders, sea gulls, and large catfishes. Tertiary carnivores include fin fishes like perches, tarpons; birds like eagles; sharks etc. Fish is considered as an important consumer in the mangrove ecosystem predating algae, zooplankton, isopods, amphipods, crustaceans, pelecypods, mollusks, prawn and other shell fish and fin fishes (Rooker 1995; Colombini et al. 1996; Williamson et al. 1994). Decomposers are the organisms which produce simple basic elements as their food from dead or decomposed organic matter. They include fungi, bacteria, protozoa etc. Food Chain is the passage of energy materials from producers through a sequence of herbivores and a number of carnivores. Food web is a group of interrelated food chains.

9.5 Detritus Food Web

Detritus is the material resulting from the disintegration of dead organic remains. The detritus food webs in the Sunderbans are based on the phytoplankton, zooplanktons, sea grasses and mangroves, benthic and epiphytic algae including salt marshes which are the most important primary producers in this estuarine system. Much of the primary production in mangrove ecosystems is not consumed to any significant degree by herbivores that directly utilize the dominant plants for food appear to be a relatively minor component. It is thus the enormous production of plant detritus that provides the energy to drive the biological machinery of the mangrove ecosystems. Primary detritus is derived from the mangrove litters. Mangrove litters are exported to adjacent systems by tidal waters. A major part undergoes decomposition because this litter contains lignin and cellulose, which cannot be digested by macro-consumers.

9.6 Mangrove Litter

Mangrove litter fall are seen the heaviest during the dry season. Mangrove leaves, barks, twigs, branches, flowers, fruits and seeds compose the mangrove litter. Mangrove litter comprises of about 80–90 % mangrove leaves that undergoes the decomposition process and enter in to the mangrove ecosystems. Mangrove litter fall are measured and the content of the litter is analysed. The range of the volume of the litter fall varies from species to species, habitat to habitat depending on the seasonal variations (Imbart and Menard 1997; Saenger 1998; Conacher et al. 1996). Detritus contributes about 60 % of the primary production in the mangrove ecosystems.

Litter of the Sunderbans mangrove forest consist of both vegetative parts like leaves, barks and twigs and reproductive parts like flowers and fruits which are accumulated on the forest floor in different geomorphic situations like river flood plains, islands and river banks. It is observed that the reproductive materials like flowers, fruits and seeds are comprised of about 10 % of the litters for only the mangrove species, *Agiceros majus*. Litter production of *Avicennia marina* and *Excoecaria agallocha* is high during post-monsoon period (Figs. 9.3 and 9.4).

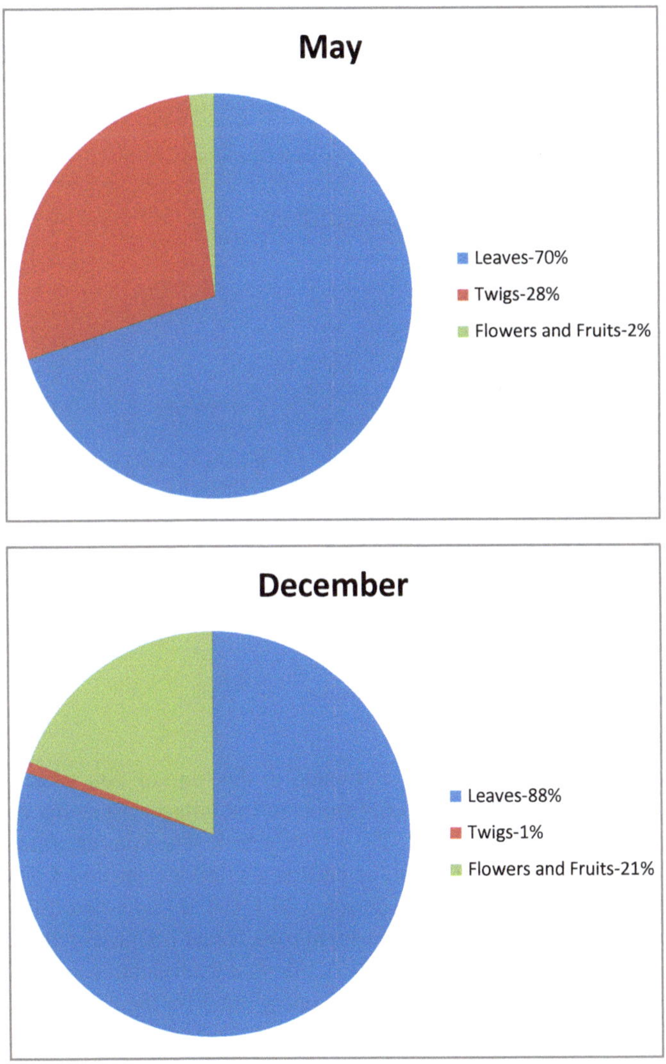

Fig. 9.3 Litter components of *Avicennia marina*

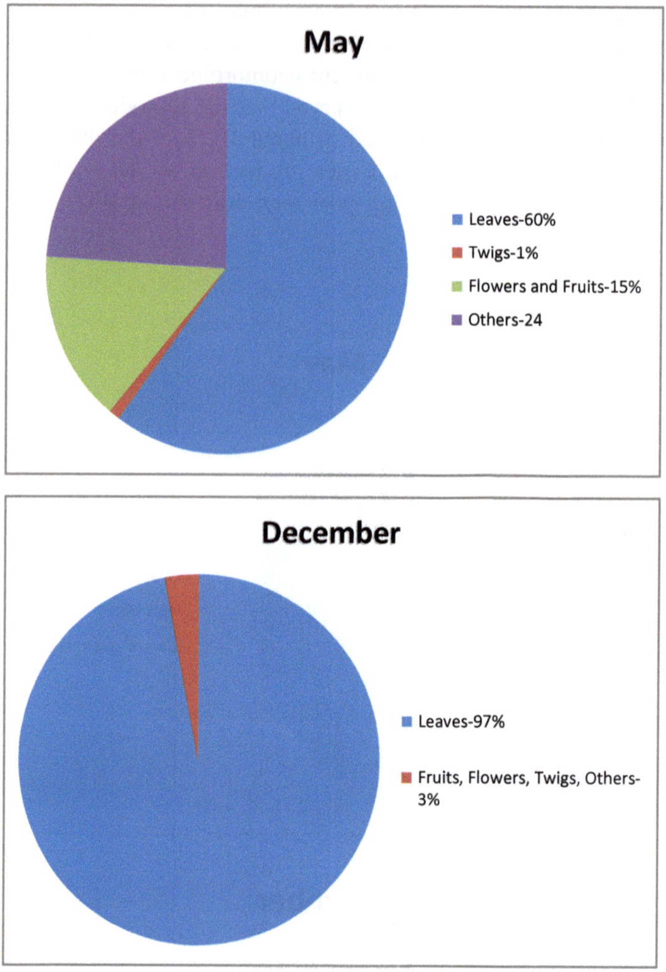

Fig. 9.4 Litter components of *Excoecaria agalocha*

Average litter fall of mangroves is estimated to the tune of 900 g C/m/year. Ten to fifteen percent of this litter is transported to the adjacent systems by the tidal currents and wave actions. Further 250 g C/m/year leaches out as soluble organic matter and 200 g C/m/year still exists as particulate organic detritus. Loss in carbon content (40–70 %) and increase in nitrogen (0.1–0.4 %) as a result of microbial immobilization of mangrove leaves in mangrove forest litters enrich the nutritive content in the aquatic system (Das 2010). Abundant production of plant detritus from the mangrove forest litter provides the energy to drive the living systems of the mangrove ecosystems.

9.6.1 Break Down of Mangrove Litters

Mangrove litters consisting of vegetative and reproductive structures undergo decomposition as detritus. Both the physical and biological processes accelerate the break down mechanisms of the mangrove litters. Physical processes are controlled by the tides, terrestrial run off and rainfall and the biological processes are controlled through decomposition, mineral uptake and faunal activities. These processes control the rate of nutrient import, export and storage in this dynamic ecosystem. The result of these flow and transformation is the import of organic matter from terrestrial ecosystems and export of both particulate and dissolved organic matter to adjacent marine ecosystems.

The aquatic organisms depend upon the detritus derived from the decomposition of mangrove forest litters as their primary food source. Mangrove litters are drifted partially to the adjacent marine ecosystems at the rate of 2–14 % of the entire litter produce by the tides and currents. Organic detritus exported directly from the mangroves provide food and habitat for the aquatic organisms like prawn and fishes (Primavera 1998; Daniel and Robertson 1990; Robertson and Blaber 1992). Macro-consumers can not digest the major part of the mangrove litters as they contain lignin and cellulose and for that reason maximum content of mangrove litters undergoes decomposition. It is observed that mangroves litters of Sunderbans undergo three kinds of break down although these mangroves litters decompose slowly. These types of break down are- (i) leaching, (ii) attacking by fungi and bacteria and (iii) mechanical fragmentation. Leaching occurs on the soluble organics into surrounding sediments and waters. Fungi and bacteria attack on soluble and particulate organic matter. Mechanical fragmentation happens due to movement and feeding of the animals and tidal and wave actions. This type of break down reduces the size of particles in particulate organic matter and exposes the organic matters. Nutrient enrichment of mangroves leaves results from loss in carbon content and in increase in nutrient export to adjacent marine ecosystems is reported.

9.6.2 Decomposition Process

Mangrove vegetations are the principal source for the produce of mangrove forest litters that contribute the major portion of primary production in the mangrove ecosystems after their degradation by the microbial activities. Aquatic organisms in the mangrove ecosystems of the Sunderbans depend on the detritus derived from the decomposition of the mangrove forest litters. The mechanism of detritus production and transformations of organic matters from these litters are very lengthy and complex processes. Mangrove litters decompose very slowly. Generally leaching of soluble organics into surrounding waters and sediments, fungal and bacterial attack on soluble and particulate organic matter and mechanical fragmentation due to tidal and wave action are the ways of breakdown of the mangrove litters. Fungi, bacteria and protozoa are the agents that accelerate the process of decomposition.

Table 9.2 Observances of sequential changes of mangrove leaves during the process of decomposition in the laboratory

Phase	Observations	Periodical sequence
1.	Mangrove leaf colour becomes pale	3 days
2.	Black spots are seen on the leaf	6 days
3.	Fungal mycelia are observed on the leaf	10 days
4.	Change of leaf bottom into dark black	15 days
5.	Leaf becomes tender and soft	16 days
6.	Decomposition of complete leaf tissues	25 days
7.	Decomposed leaf shows inner venation	27 days

The greenish look of the mangrove leaves is abolished within 3 days as observed in the laboratory experiment. The mangroves leaves are kept in the Petridis filled with mangrove sediments with sufficient brackish waters collected from the rivers of the Sunderbans. Black spots are occurred on the leaves after 6 days and spores are visible under the microscope. As a result of the fungal attack on the mangrove litters fungal mycelia dominates over the leaves after 10 days and black colour appears at the bottom side that lead to the leaves becoming soft and tender after 15 days. Exposure of inner venation after complete decomposition due to degradation of tissues is observed after 25–27 days (Table 9.2). This process of decomposition observed in the laboratory situation might be accelerated in nature and take less period for the mechanical fragmentation of mangrove litters due to tidal and wave actions.

In the mangrove ecosystems the decomposers are continuously grazed by the protozoans during microbial degradations, fungi and bacteria utilize the resistant parts and improve protein content of the detritus. This leads to the great potential food value with a rich protozoa-bacteria-detritus system. Macro-consumers may ingest this food complex within this system. The detritus becomes poorly nutritive with time. The detritus play the key role enriching the primary food source in the mangrove ecosystem that feeds the aquatic organisms of both primary and secondary levels.

In the process of decomposition bacteria and fungi attack the mangrove forest litter. Size of the particulate materials is reduced by mechanical fragmentation due to biotic and abiotic factors. Then bacteria and fungi utilize the resistance parts and improve protein content of the detritus. The decomposers are continuously grazed by protozoans during microbial degradation. This creates a rich protozoa-bacteria-detritus system with great potential food value. The complex may be ingested by larger consumers like fish or gastropods at the duration of the intervals.

Detail characterisation of the input and output of minerals in the process of decomposition and their physico-chemical natures of this dynamic ecosystem is essential that help in the understanding the process of decomposition and production of detritus from the mangrove litters.

9.6.3 Microbial Utilizations

The microbial portion of the detritus food web which involves detritus production and other transformations of organic matter are more complex and consume more energy and materials than the macro-heterotrophs. Generally detritus particles which are ingested are smaller than 0.2 mm in diameter and often appeared to be aggregates of much smaller particles. These particles are covered by large numbers of microbes protozoans and tiny metazoans. The nourishment for detritus feeders from bacteria comes from bacteria and fungi involved in decomposition rather than detritus.

Bacteria are useful as food in two ways – (i) directly as nourishment and (ii) assist organisms in digestion. Detritus is initially colonized by bacteria with later colonization by fungi algae ciliates flagellates and nematodes. Bacterial and fungal degradation of organic detritus is therefore accompanied by bacterial grazing by biota, where the whole microbial community is grazed by macrofaunal detritus feeders like fin fish and shell fishes. These macrofaunal grazing activities on microbial communities stimulate microbial growth and detrital mineralization. It is recorded that detritus feeders assimilate organic detritus poorly, but efficiently digest the microbial community living on the detritus. The major sink of soluble organic matter appears to be microbial utilization as – (i) microbes are efficient to take soluble organic matter even in very low concentrations and (ii) they develop at sites of release of dissolved organic matter so that much uptake is closely coupled with release. Distribution of the juvenile forms of the prawns and their abundance in the mangroves are strongly influenced by the detritus-rich muddy substrate in the mangal (Vance 1992; Rajendran 1997; Mohan and Siddeek 1996).

9.7 Nutrient Recycling

Loss in the carbon content and increase in nitrogen concentration for microbial nitrogen immobilization result nutritive enrichment of mangrove leaves. Both the nutrient input from terrestrial sources and nutrient export to adjacent marine ecosystems is observed, as mangrove ecosystems are open systems. Sediments reworking by macro-consumers brings limiting nutrients to surface for primary production. The animal- sediment relationship provides new habitats for other organisms, alters water turbidity, affects production and provides aeration. Thus macro-consumers influence nutrient cycles through (i) direct regeneration of nutrients, (ii) mastication enhancement and packaging of detritus and (iii) reworking of sediments.

Table 9.3 The rate of nutrient inputs and exports

N as NO_2	Input	0.3 kg/day	N flux as NO_2	23 kg/day
N as NO_3	Input	4.0 kg/day	N flux as NO_3	3 kg/day
P as PO_4	Input	0.5 kg/day		
Si as SiO_2	Input	370 kg/day	Si flux as SiO_2	3,100 kg/day
N as total nitrogen	Input	70 kg/day		
P as total phosphorus	Input	32 kg/day		
C as total organic carbon	Input	700 kg/day		

9.8 Nutrient Inputs and Exports

Detail characterization of the input and output of minerals and the physico-chemical natures of this dynamic ecosystem is essential that helps in understanding the process of decomposition and production of detritus from the mangrove litters. Nutrient input from terrestrial sources and nutrient exports to adjacent marine ecosystems are observed as the mangrove ecosystems in the Sunderbans are open systems. The nutrient inputs and exports through nitrogen phosphorus silica carbon etc. are important for this open type of ecosystem (Table 9.3).

9.9 Energy Flow

Energy flows through different pathways in the food web of mangrove ecosystem. The organisms involved in this food web are almost a group of detritus consumers namely herbivorous and carnivorous crustaceans, amphipods, molluscs, insect larvae, polychaetes, nematodes, fin fish and shell fishes (Fig. 9.5). The mangrove detritus is the principal source for input of energy in this ecosystem. Micro and macro algae, salt marsh, sea grasses and mangroves are the producers in this ecosystem. The active microbial community present in the sediments accomplishes nutrient recycling in the mangrove ecosystem through the microbial decomposition. The rate of recycling of nutrients is inversely proportional to body size. Herbivorous organisms do not play significant role in the energy flow of mangrove ecosystem (Fig. 9.6). Transfer of energy flow is comparatively less from the primary production to herbivorous tropical level. Poikilothermous animals are mainly comprised of the community of herbivores like crustaceans, crabs, fishes etc. Herbivores like homeothermic organisms (eg. Cattle, goats, deer, birds etc.) are less present to consume primary production (Table 9.2).

Omnivorous organisms are present in the mangrove food web as because all animals might have special requirement with different foods of both quality and quantity in space and time (Fig. 9.7).

Mangrove consumers (eg. Crabs, mullets, snails) influence nutrient cycles through direct regeneration of nutrients by grinding the ingested materials and

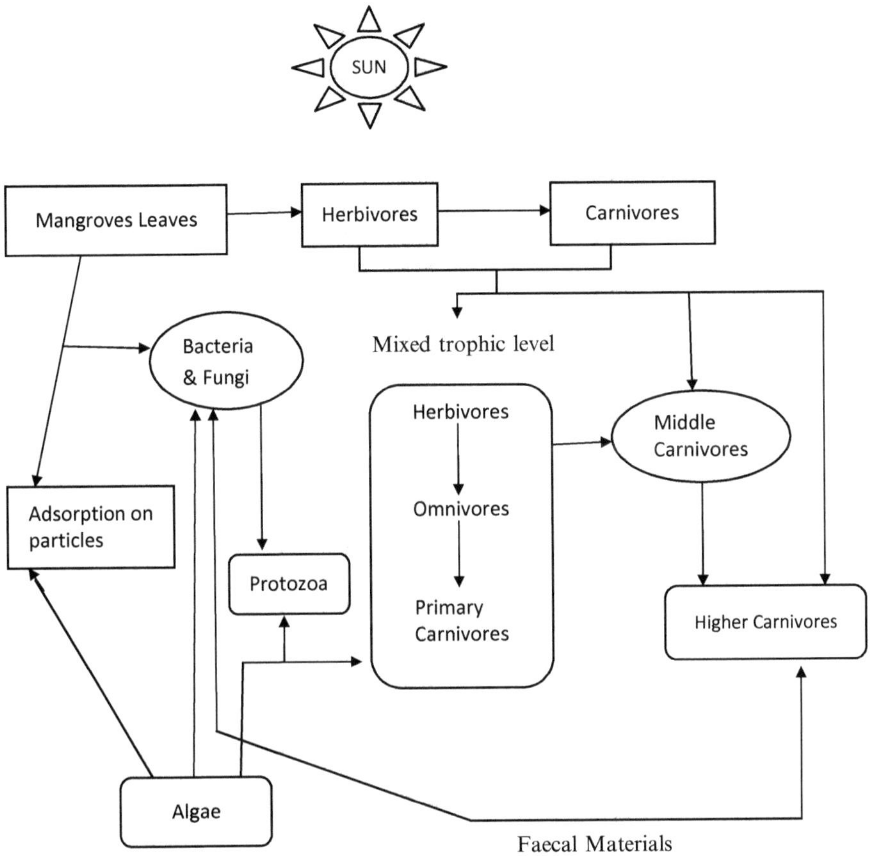

Fig. 9.5 Compartment model of food chain of Sunderbans mangroves ecosystems

accelerates decomposition of egested materials by providing new surfaces for microbial attack. The microbial communities are very important in nutrient regeneration than macroconsumers.

9.10 Functions of Mangrove Ecosystem

The functions of mangrove ecosystems are classified into three categories: hydrological, biogeochemical and ecological functions. Hydrological functions are mainly related to the flood water retention through surface and ground water recharge that has the socio-economic benefits like alternative food protection, reducing damage to property and natural resources by flood water storage, maintenance of habitat and water supply. The biogeochemical functions include nutrient retention through

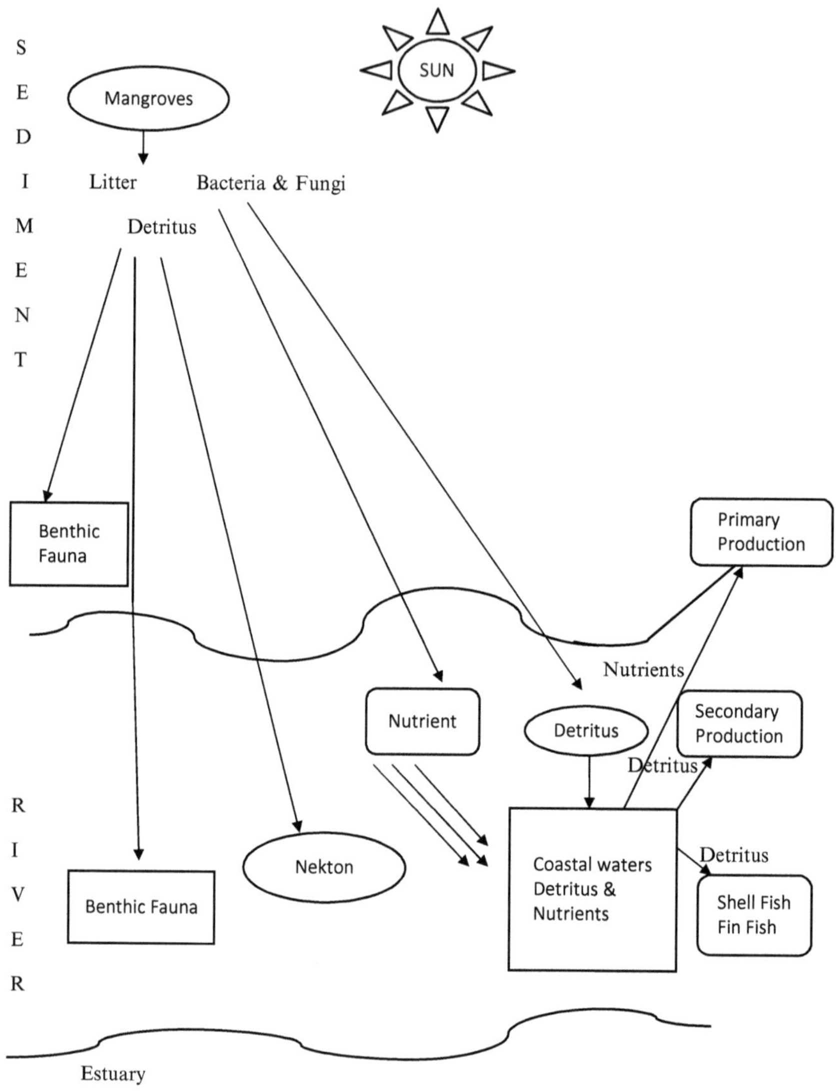

Fig. 9.6 Energy flow in mangrove ecosystem

export and salinity retention in water and soil which have the benefits of waste disposal and improvement of water quality. The ecological function involves in the nursery and habitat structure via nursing of plants, animals and microorganisms, maintaining biodiversity, landscapes and structural diversity. These functions have the direct impact on the socio-economic benefits like tourism, recreational activities,

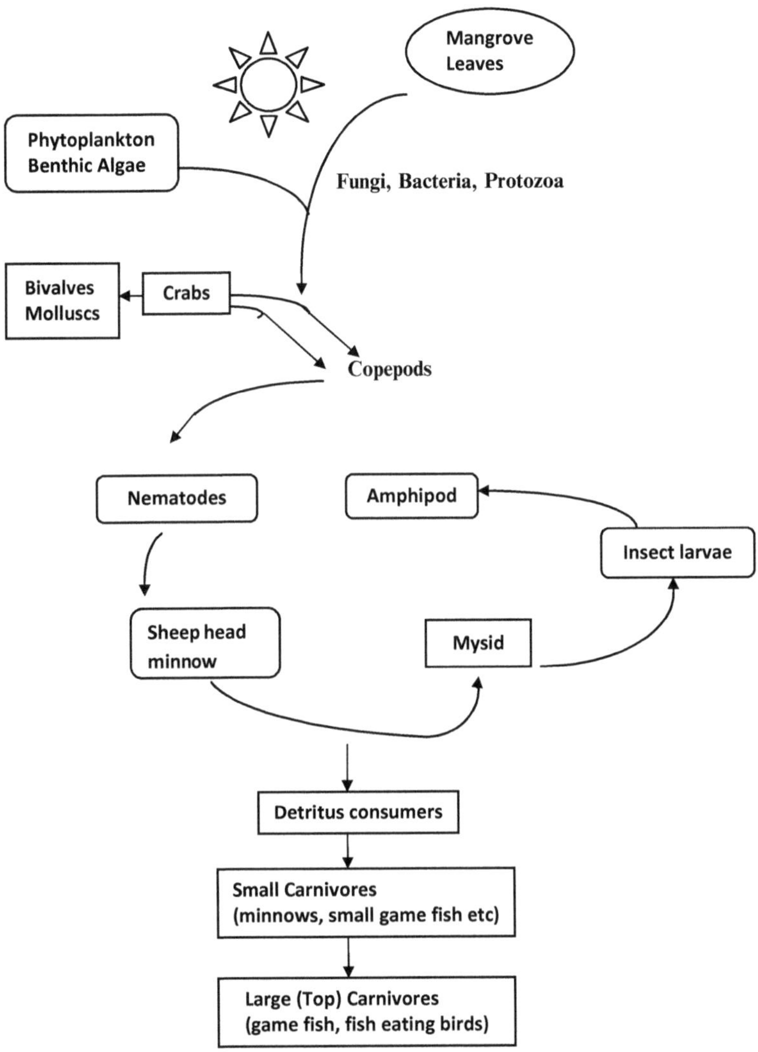

Fig. 9.7 Typical detritus food chain in the mangrove ecosystem at Sunderbans

fishing and aquaculture. Other than these mangroves help in the formation of soil through trapping and mineralisation of debris. Mangroves ecosystem enhances in sorting the nutrient from the rich organic materials by the tidal actions through river systems and the wave action of the coastal sea. This ecosystem is the shelter for fish, marine invertebrates, mollusca and birds that enrich the productivity of the ecosystem.

9.11 Mangrove and Man

The mangrove ecosystem is really splendid to observe that *Ceriops* and *Excoccaria* stand at the confluence of river mouth whereas *Phoenix* and *Nypa* are in the bank of creeks, tidal inlets and rivers having with different stages of diversity in a quiet environment of grand symphony of silence (Vanucci 1989). Here hundreds of stamina of *Sonneratia* flower are exposed at night to tempt the bats for process pollination and the same process is used to do by the swarm of bees in the *Agiceros* sp. Bee makes their nest on the branches of *Agiceros* sp for the production of honey and wax. Monkey covers its body with layer of mud before going to taste the honey in disguise.

The inhabitants of Sunderbans know it well how to construct the indigenous boats using the mangrove trees specially with the *Avicennia* for the keel, *Ceriops* for knees, *Rhizophora* for the mast and *Xylocarpus* for the plank for the purpose of boat making. *Rhizophora* and *Brugeira* have the power of tolerance in the high humidity whereas the *Avicennia* has another power of tolerance in high salinity. *Rhizophora* is known for the best charcoal with high calorific value without any smoke.

Fishermen, honey-collectors, Loggers, Hunters and even the pirates along with the luxuriant mangroves and marsh make this mangrove ecosystem a unity in diversity. This ecosystem is dying of for over exploitation. But the mangrove dwellers must be the presence of this ecosystem through aqua-agri-silvi-culture systems that fulfill the requirements for basic human needs.

9.12 Summary

Mangrove vegetation on the river flood plains has been reducing year by year because of reclamation for aquacultural practices. Indiscriminate cutting of the mangrove forest for fuel wood, timber and branches for using country house construction and preparation of frame for beetle leaf cultivation is another type of threat for the existence of the mangroves. Some important species of mangroves that only grow in a particular physico-chemical environment on the verge of extinction due to conversion of land use pattern for agriculture, aquaculture and rural residential development setting up for the island-dwellers in the Sunderbans. At present 54 islands out of total 102 islands are already reclaimed and turned into the residential villages only in a span of 100 years. This reclamation of the mangrove forests and conversion of land use pattern ultimately results in the disturbance of entire mangrove ecosystem. The Sunderbans mangroves facing the point and non-point pollution near the estuarine river mouth areas is another threatening factors as this halophytic vegetation is vulnerable to the biodegradable and stable compounds thrown and transported from the land sources.

References

Colombini I, Berti R, Nocita A, Cheelazzi L (1996) Foraging strategy of the mud skipper Periophthalmus sobrinus Eggert in a Kenyan mangrove. J Exp Mar Biol Ecol 197(2):219–235

Conacher CA, O'Brien C, Horrocks JL, Kenyon RK (1996) Litter production and accumulation in stressed mangrove communities in the Embley river estuary, north eastern gulf of Carpentaria. Aust Mar Freshw Resour 47:737–743

Daniel PA, Robertson AI (1990) Epibenthos of mangrove waterways and open embayments: community structure and the relationship between exported mangrove detritus and epifaunal standing stocks. Estuar Coast Shelf Sci 31(5):599–619

Das GK (2010) Biodiversity in Sunderbans. In: Dwivedi AK, Tripathi SC (eds) Environment and biodiversity, environmental pollution and risk to biodiversity. Lambert Academic Publishing AG & Co., Saarbrucken, pp 1–34

Farnsworth EJ, Ellison AM (1996) Scale dependent spatial and temporal variability in biogeography of mangrove root epibiont communities. Ecol Monogr 66(1):45–66

Imbart D, Menard S (1997) Structure de la vegetation el production primaire dans la mangrove de la Baie de Fort de France. Martinique (FWI). Biotropica 29(4):413–426

Mohan R, Siddeek MSM (1996) Marine habitat preference, distribution and growth of post larvae, juvenile and pre-adult Indian white shrimp, Penaeus indicus H. Milne Edwards. In Ghubat Hasish Bay, Gulf of Masirah, Sultanate of Oman. Fish Manage Ecol 3(2):165–174

Phillips A, Lambert G, Granger JE, Steinke TD (1994) Horizontal zonation of epiphytic algae associated with Avicennia marina (Forssk.) Vierh pneumatophores at Beachwood Mangroves. Nature reserve. Durban. South Africa. Bot Mar 39(2):167–175

Primavera JH (1998) Mangrove as nurseries: shrim population in mangrove and non-mangrove habitats. Estuar Coast Shelf Sci 46(3):457–464

Rajendran N (1997) Studies on mangrove associated prawn seed resources of the Pichavaram, south east coast of India. Ph. D. thesis. Annamalai University, India

Robertson AI, Blaber SJM (1992) Plankton epibenthos and fish communities. In: Robertson AI, Alongi DM (eds) Tropical mangrove ecosystems. American Geophysical Union, Washington, DC, pp 173–224

Rooker JR (1995) Feeding ecology of schoolmaster snapper, Lutjanus apodus (Walbaum) from south western Puerto Rico. Bull Mar Sci 56(3):881–894

Saenger P (1998) Mangrove vegetation: an evolutionary perspective. Mar Freshw Res 49(4):277–286

Vance DJ (1992) Activity patterns of juvenile penaeid prawns in response to artificial tidal and day-night cycles: a comparison of three species. Mar Ecol Prog Ser 87:215–226

Vannucci M (1989) The mangroves and us; a synthesis of insights. Indian Association for the Advancement of Science, New Delhi

Williamson I, King C, Mather PB (1994) A comparison of fish communities in unmodified and modified inshore habitats of Raby Bay, Queensland. Estuar Coast Shelf Sci 39(4):401–411

Chapter 10
Anthropogenic Interferences

Abstract The vibrant ecosystem of Sunderbans with its lush green mangrove forest, various types of aquatic species including many rare endemic flora and fauna, birds, crocodiles and the famous Royal Bengal Tiger is a unique Biosphere Reserve. Unfortunately such a productive ecosystem is now exposed to threats of extinction. Besides the natural global phenomena, the land degrading anthropogenic activity is one of the prime factors for this ecological decline. In a forest-society interface, unplanned over exploitation of natural resources is very common. In the case of Sunderbans also, as a result of continuously increasing population pressure almost half of the mangrove forest have been cut down to supply fuel wood, land reclamation for settlement and aquaculture and various other purposes. The situation has become precarious as there seems to be a striking imbalance between exploitation and replenishment, thereby making the ecosystem most fragile. Indiscriminate prawn seed collection, refuse and sewerage discharge from urban areas, spillage of toxic pollutants as well as trashes in the tourist spots of Sunderbans are other anthropogenic factors responsible for environmental degradation of the area. Suitable strategies for the conservation and maintenance of optimum ecological condition of this unique biodiversity have to be planned involving local people, Government and other social organizations.

Keywords Sunderbans • Mangroves • Prawn seeds • Shell fish • Fin fish • Bidyadhari river • Royal Bengal tiger

The rich resources of the estuarine and coastal waters are under tremendous pressure not only because of polluted sewage thrown by the man made canal into the Bidyadhari River waters, but for extensive overfishing in the river waters of the Sunderbans. The capture of the post larval penaeid prawns that are potentially quite high and commercially exploitable from the salt water courses of the deltaic Sunderbans even just a decade ago are presently showing a dwindling trend. The fishing pattern of the post larval stage of penaeid prawns along the marginal waters of salt water courses of the deltaic Sunderbans caused further hazards in the form of bank erosion and impediment in nutrient recycling. The fringy mangrove belts which act as the chief source of nutrient supply to these water bodies thus seem to be severally affected both by natural and human activities.

G.K. Das, *Estuarine Morphodynamics of the Sunderbans*,
Coastal Research Library 11, DOI 10.1007/978-3-319-11343-2_10

In addition to that the domestic sewage of Kolkata metropolis and the industrial effluents from East Kolkata released on the regular basis in to the river waters of the Bidyadhari through different canals that ultimately flows to the Sunderbans mangroves zones is the other problem. Mangroves are sensitive to these different types of pollutants which are significant threats to the mangrove ecosystems. Mangroves are further sensitive to oil spills released by the mechanized fishing boats. Vegetations of the Sunderbans are found to be susceptible to pollution in the form of floating oil, as because it deposits on mangrove roots when the tide drops and contaminates the sediments. Mangroves suffer from suffocating or toxic effects of the oil, or from the dwelling of burrowing animals. The incidence of tiger straying almost regularly for the gradual decreasing of the forest land including Phoenix bushes and the dense jungles, the unique denizen for the Royal Bengal tigers is another effect of the anthropological interferences.

10.1 Prawn Captures

Salt water courses of the deltaic Sunderbans are zones of high potential so far as captures of prawn seeds are concerned. However, during last couple of years the dwindling trend is observed regarding the prawn seed capture in this zone. The fishing pattern of the post-larvae stage of penaeid prawns along the marginal waters of the rivers of the Sunderbans caused further hazards in the form of bank erosion and impediment in nutrient cycling. The situation is further aggravated as large numbers of brackish and marine fishes are encountered in this process along with the rampant exploitation of the prawn seeds. Geomorphic mapping of the river catchment areas of the Sunderbans has been done on the basis of the toposheets of the Survey of India. All the physical and chemical parameters of the soil and waters are estimated following standard methods. Statistical surveys of the collection pattern of the prawn seeds are undertaken from different collection centers of Sunderbans. Rising scarcity of prawn seeds leads to sharp increase towards expenditure per person for harvesting of the 1,000 seeds per day. There is a decreasing trend in the rate of mortality of the prawn seeds. With passage of time number of seeds transported to different fisheries has also been depleted considerably, sometimes to the tune of tenfold as evidenced from the data of 1998 to 2010. The proliferation of the prawn seeds requires favourable physico-chemical, compositional and textural parameters of sediments, geomorphology of the drainage system and tidal cycles within them. Data reveal that the destruction of the encountered species disturbs a bulk of the population of the coastal water biotic community and as a consequence brings about an imbalance in the food webs on the water dwelling biota and enormous stress because of rampant exploitation.

The mangrove belts of Sunderbans which act as the chief source of nutrient supply to those water bodies also get badly affected by artificial means of fishing in addition to natural hazards arising from meteoric storms and cyclones. The author has tried to identify and delineate the geomorphic situations and other physico-chemical environments that control the prolific occurrence of the post-larvae of prawns in the rivers of Sunderbans which are later used for culture in brackish water coastal fisheries. Certain management measures have been proposed to safeguard the disaster following approaches related to morphological and hydrodynamic appraisement.

10.2 Environmental Requirements for Habitat

- Salinity ranges between 10 and 25 parts per thousand (ppt).
- Optimum temperature range for good growth and survival rates of tiger prawn is 25–30 °C.
- Minimum acceptable Dissolved oxygen (DO) level is 3–4 mg/L.
- Water pH of 5 and below is lethal to prawns like tiger prawns. pH range of 7–8.5 is optimal for growth and development of prawn.
- Presence of Hydrogen Sulphides (H_2S) and low Dissolved Oxygen (DO) level are detrimental to their survival as tiger prawns are the bottom dwellers (Anonymous 1993).

10.2.1 Sediment Composition

The penaeid prawns are detritus feeder and so the detritus inherently present in the sediments has a vital role for their growth and proliferation. The detritus materials are recycled through physical and biological processes. Compositionally, the sediments collected from the river flood plains composed of both lithogenic (85–90 %) and biogenic (10–15 %) components and dominantly made up of quartz, mica, rock fragments and clay minerals. Texturally sediments are mostly silty (90 %) with subordinate fine sand and clay constituting the rest (Table 6.5).

10.2.2 Physico-chemical Control for Sustainability of Prawn Seeds

The physico-chemical parameters like salinity, water depth, light penetration, pH, total organic matter, nutrient concentration (NPK) etc. control the growth and proliferation of penaeid prawn seeds. The study area typically belongs to the tropics with average maximum temperature of winter and summer months ranging from 20 °C to 35 °C respectively, the annual rain fall between 1,900 and 2,100 mm. The sediment salinity of the study area of Sunderbans ranges between 4.93 and 5.8 ppt and the pH is slightly alkaline ranging between 8.0 and 8.3. Chemical analyses of the collected sediments samples reveal that the available nitrogen (Av.N) ranges from 880 to 1,540 kg/ha, available phosphorus (Av.P) varies from 11.4 to 26.0 kg/ha and available potash (Av.K) ranges from 2,083 to 2,957 kg/ha. The organic Carbon shows a range between 0.38 and 0.69 % and the total organic matter in sediments ranges from 0.66 to 1.19 % (Table 4.3). It is found that the ambient environment supplies the plenty of nutrients that requires for the natural growth of mangroves of the Sunderbans. Water salinity varied from 18 to 25 ppt as compared to the water temperature 18–21 °C during post-monsoon session (Table 3.1). The pH values remained almost constant ranging from 7.3 to 7.7. The dissolved oxygen (DO) content varied from 3.0 to 7.2 mg/L (Table 3.1). Monsoon season shows a marked difference in the hydrological parameters from other seasons due to heavy rainfall and land run-off.

10.3 Geo-hydrological Control Related to Capture

Geo- hydrology of the rivers of Sunderbans controls the density of occurrence of prawn broods and the related fishing methods practiced along the salt water courses of the deltaic Sunderbans. Geomorphology of the drainage systems has tremendous role over population density of penaeid prawns in specific zones of the typically meandering saline or brackish water courses. The longitudinal flood and ebb-flows and the transverse helicoidal flows in meandering systems also add to the drifting behaviour of the prawn seeds (Fig. 10.1). The complicated process of hydrodynamics and sediment supply in meandering water courses regulates deposition of point bars on the inner convex banks erosion scarps on the outer concave sides.

The phenomena of undercutting on the outer concave banks and subsequent collapsing bring about a constant modification of the river morphology by laterally shifting its course. On the other hand, mangrove forests commonly fringe the natural levees and the upper reaches of intertidal point bars of the river and help stabilization of the river courses. The seasonal shading off litters is carrying out constant recharging of nutrients in the water by the mangrove plants. Field observations reveal that areas of marginal waters up to a depth of 2 m along the erosional, outer concave banks of the rivers are more suitable for the potential occurrence of the penaeid larvae. The association of mangrove forests in the nearby areas plays a significant role contributing the nutrients in the ambient waters.

10.4 Seed Availability

Thakuran, Jhilla, Bidya, Saptamukhi rivers yield lower percentage (64–69 %) of shell fish seeds, mullets (7–8 %) and thread fins (3–5 %). The contribution of miscellaneous species is naturally higher (15–24 %) in these river estuaries. The annual average of fin and shell fish seed catch is 10,077 nos./net/day of which 65–75 % is comprised of shell fish, 3–5 % tiger shrimp and 20–30 % fin fishes. The peak period of seed availability particularly for the period from June to September recording average total catch between 9,750 and 19,110 nos./net/day of which tiger

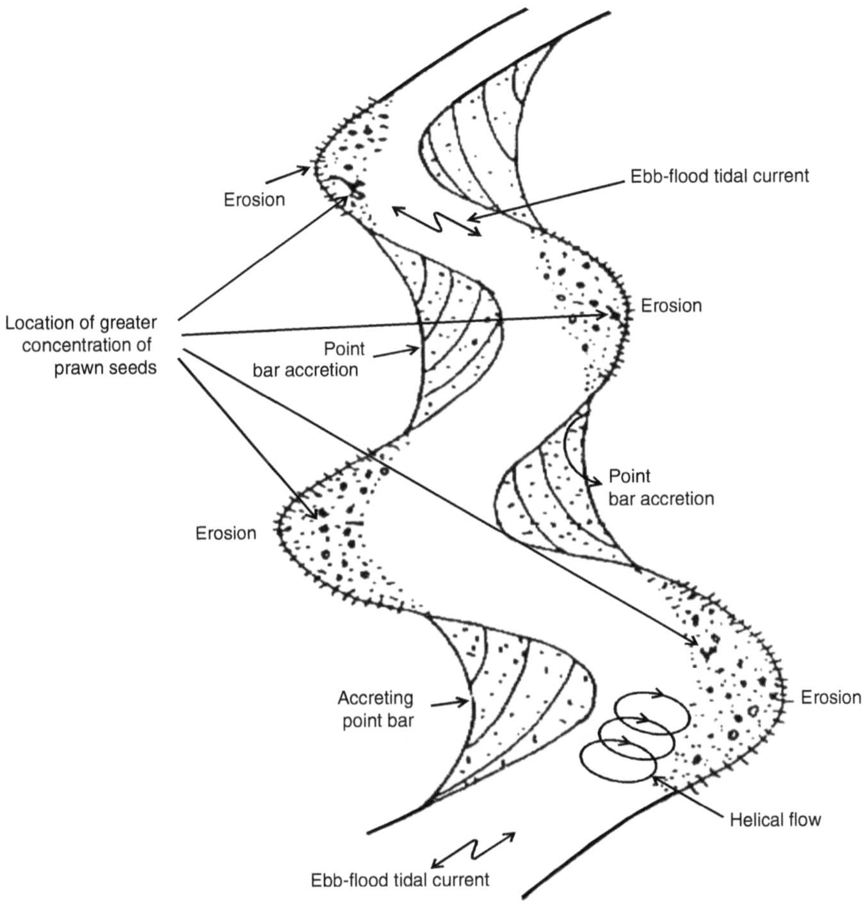

Fig. 10.1 Meandering course of river showing erosion and accretion sites and related concentration of prawn seeds

prawn between 377 and 960 nos./net/day. Availability of seeds remains almost constant in the rest period of the year in the estuarine Sunderbans.

10.5 Seed Composition

The seeds available in the river waters are composed of fin and shell fish seeds in the estuarine Sunderbans (Table 10.1). About 72 % shell fish species, 10 % mullets and 7 % thread fins are the principal constituents of the seed composition. The other species contributing the total seeds collected from the system are *T. ilisha* (0.3 %), *L. calcarifer* (0.9 %), *M. gulio* (1 %), *H. neherius* (1 %), *T. jarbua* (0.5 %) and miscellaneous (8 %).

Table 10.1 List of the encountered species collected from the trash after sorting from the river waters of Sunderbans

Fin fish	Shell fish
Setipinna phasa	*Metapeneus monoceros*
Corica soborna	*Metapeneus lysianassa*
Anguilla bengalensia	*Metapeneus mogiensis*
Reconda rusheliana	*Parapeneopsis stylifera*
Cynoglossus arel	*Palaemon* (*Exopalaemon*)
Liza tade	*Stylitera*
Leiognathus daura	*Alphaeus* Sp.
Pellonaditchela	*Acetes* Sp.
Ophisthopterus tardoore	
Coilia dussumiery	
Hilsa ilisha	
Stolephorus bucaneeri	
Neenchelys buitendijki	
Strongylura strongylura	
Trypanchenichthys suratensis	

10.6 Drifting Pattern

The post larvae of prawn swarms appear in coastal waters and move shoreward up to 2 weeks in their post larval stage. They continue migration towards mangrove swamps and other brackish water salt marshes, which serve as their nurseries or feeding grounds (Das and Bhattacharya 1999a). The species are omnivorous and feed at the bottom of the muddy swamps and marshes. This is evidenced by the presence of very minute vegetable and animal matters and detritus constituents in their guts. Because of their detritivorous feeding behaviour the seeds occur abundantly along margins of the river point bars often girdled by fringing mangroves. The point bars are generally accretionary where loose sedimentary particles are reworked by every tidal cycle. The typical helical flow in the meandering channel systems help drifting the prawn seeds from these accretionary zones towards the opposite erosional portions of the meander bends (Das and Bhattacharya 1999b). Being planktonic, the seeds are also drifted by the dominating tidal currents.

10.7 Different Phases of Prawn Seeds (*P. monodon*)

Spawning: Mating of Penaed soon after molting is occurred. As the mature female lays the eggs, the spermatophores stored in the seminal vesicle are simultaneously released and the eggs thus fertilized.

10.7.1 Metamorphosis of Prawn

Nauplius: The eggs of *P monodon* hatch into nauplius within about 13–14 h after spawning, which feed on the yolk, stored in the body and do not feed on an external diet.

Zoea: *Nauplius* transform in to zoea in 36–37 h after hatching. They feed on planktonic larvae. The body length of zoea varies between 0.092 and 2.24 mm from 1st to 3rd stage.

Mysis: When zoea molts for the 3rd time and changes into mysis, the morphology resembles that of adult prawns. They feed on planktonic larvae and then transform into the post larval stage.

Post-larvae: Mysis molt three times in 3 days and transform into the post larval stage. When the post larvae molt 10–12 times, they are able to move on sand like the adult prawns. Again, larvae, which have molted about 20 times, are called juvenile prawns.

10.8 Collection Method

Generally drag net is used in the lower reaches of the Sunderbans for *P. monodon* fry collection. It is a rectangular net of length 3.0 m and width 1.5 m, made of fine nylon net clothes. All four ends of net are firmly stitched around a rectangular split bamboo frame (1.25 × 0.75 m). Both ends of the net are attached to a long nylon rope and one-person generally teenage girls holding both nylon ropes together and dragging the net from behind and operate the net in the shallow water areas. The nets are generally operated in waist-to chest deep water. Sometimes prawn seeds are collected through stationary gears by operating nets from the country boats (Fig. 10.2).

10.9 Encountered Species

The total number of encountered species belonging to the shellfish and finfish has been given in Table 10.1. Of the total important 23 species five fin fishes (*Leiognathus daura*, *Liza* sp., *Corica soborna*, *Trypanchenicthys suratensis* and *Stolephorus* sp.) and three shell fishes (*Metapeneus lysianassa*, *Metapeneus monoceros* and *Acetes* sp.) appear to be the most dominant encountered species in the present study area. Synthesis of data depicts the year wise variation from 1998 to 2010 of the dominant encountered species against the frequency of the species taking average of three seasons in a year. Data reveal that the destruction of the

Fig. 10.2 Collection of prawn seeds using stationary gears in the river waters of Sunderbans

Fig. 10.3 Sorting of prawn seeds going on after the catch

encountered species during sorting (Fig. 10.3) disturbs a bulk of the population of the coastal water biotic community and as a consequence brings about an imbalance in the food webs of the water dwelling biota. Though not very pronounced the year wise count of the most dominant species also reveals a declining trend.

10.10 Transport of Seeds

Large quantities of prawn seeds (*P. monodon*) are assembled at Nazat in North 24 Parganas District, the biggest tiger Shrimp fry market in the country from various collection centers in the Sunderbans. All the collections are transported by the mechanized boats locally called as *vatvati*. At first prawn seeds are brought in open containers to market or collection center. The traders purchase the seeds from the collectors in the fry market. Then the fisheries owners or their representatives come to purchase fry. The fry are transported to the far off fisheries after examination of the fry and negotiation of price with the traders.

Thousands of people of Sunderbans under below poverty level have found their income like daily wage system in prawn seeds collections. The Directorate of Fisheries, Govt. of West Bengal reported after their survey in 1991 that fry collectors belonging to 3,025 families in the seven blocks of Sagar, Namkhana, Kakdwip, Diamond Harbour, Kulpi, Kultali, and Canning were engaged in this profession of fry collection and annually collected 429.1 millions of prawns fry, 78.3 million of these during lean season. Of these families, more than half had a per capita annual income between Rs. 401 and Rs. 1,200. On an average they collect about 2,500 prawn seeds a day of 8 h netting operation (Anonymous 1993) in the monsoon times.

10.11 Observations

The monsoon is the best collection season for the prawn broods; pre-monsoon and post monsoon periods come next in succession. The tidal cycle also show a correlation with seasonal yield of the prawn seeds. The ebb-tidal cycles yield the maximum throughout the seasons. Again, in the first phase of flood tides during the pre and post monsoon periods and the entire spring tide period of the monsoon, the yield is quite significant. The rate of seasonal collection and the number of seeds transported to the rearing ponds are just sufficient for their commercial exploitation (Table 10.2). The following observations seem to be relevant in this regard:

- Despite application of advanced and skilled netting procedures there is a decreasing trend of capture of prawn broods per person per day (Table 10.2).
- With more and more scarcity in the collection of seeds, amount paid to persons per day per 1,000 prawn seeds harvested increases sharply (Table 10.2 and Fig. 10.4).
- There is a decreasing trend in the rate of mortality of the prawn seeds (Table 10.2 and Fig. 10.5).
- A large number of other species of fin fishes and shell fishes (Table 10.1) are also destroyed during processes of random exploitation of the prawn seeds (Das and Bhattacharya 1999b).
- Number of seeds transported to different fisheries center decreases rapidly and that also to the tune of one – tenth parts in 2010 compared to 1998 (Table 10.2).

Table 10.2 Annual variation in collection, transport, rate of mortality and amount paid for collection of prawn seeds from tidal rivers of Sunderbans

Year	Av. no. of seeds captured/ person/day	Av. no. of seeds collected/ transporter/day	Av. no. of seeds transported from station/day	Av. mortality rate (%)	Amount paid for collection of 1,000 seeds (in Rs.)
1998	225	55×10^3	90×10^3	45	106.00
1999	130	28×10^3	84×10^3	28	333.00
2000	105	23×10^3	62×10^3	22	457.00
2001	50	14×10^3	59×10^3	19	1,060.00
2002	46	9×10^3	46×10^3	16	1,210.00
2003	38	7×10^3	41×10^3	18	1,390.00
2004	33	5×10^3	36×10^3	15	1,470.00
2005	31	5×10^3	34×10^3	14	1,530.00
2006	29	4×10^3	31×10^3	15	1,590.00
2007	26	3.2×10^3	19×10^3	17	1,680.00
2008	22	2.6×10^3	18×10^3	12	1,760.00
2009	17	2×10^3	14×10^3	10	1,820.00
2010	14	1.7×10^3	12×10^3	7	1,875.00

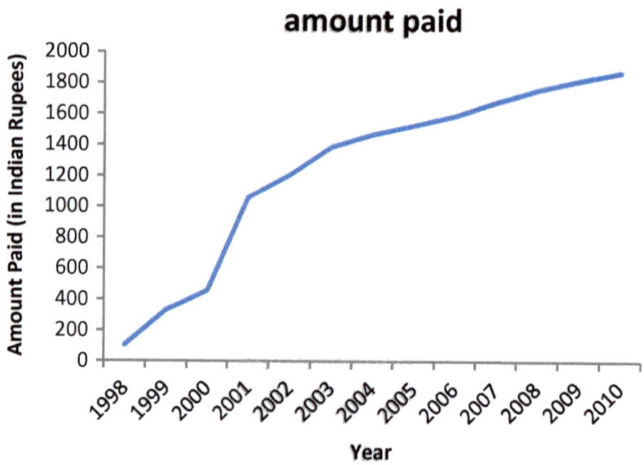

Fig. 10.4 Annual variations of amount paid to the collectors for 1,000 prawn seeds in the collection center

10.12 Management Plan

All these deleterious operations are currently continuing in most areas of Sunderbans, which, in turn, during the recent decades, have been bringing about a serious change in the geomorphology and hydrodynamics of the watercourses together with an obvious change in the nature of nutrient recycling. Again, the declining trend in the collection and rate of mortality of the prawn broods may have a direct relation

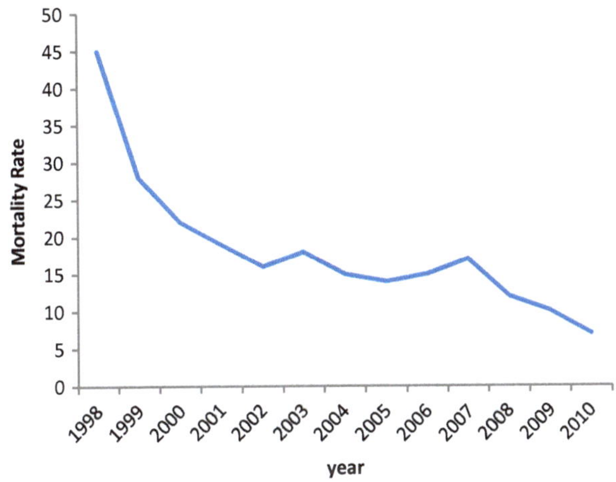

Fig. 10.5 Average mortality rate of prawn seeds during transportation to the rearing tank

with this unwise fishing method adopted in the sensitive geomorphic zones of the river systems. Thus, the author suggests management plans along to safeguard these highly productive and biologically sensitive areas.

- A minimum knowledge of salt-water course morphology should be disseminated among inhabitants.
- Data on the hydrodynamics of the watercourses must be collected prior to any attempt towards commercial exploitation.
- Systematic monitoring of the areas is badly needed for maintenance of the bank-binder mangrove saplings and marsh vegetation.
- Textural data of bed material, to the proportion of sand, silt and clay, their sorting and other mass properties should be collected in order to understand the physical properties of the substrate material.

10.13 Impact of Seed Capture

The geomorphological and geohydrological set up of meandering tidal rivers, creeks and inlets of the Sunderbans also plays a significant role in this regard. The population density of the post larva penaeid prawns and their catch or effective netting areas are mostly confined to the outer concave erosional banks having water depth of less than to the accretionary inner concave banks with point bar deposits. Nutrient recycling from the fringing forests further helps towards their proliferation. The crude fishing technique together with improper knowledge about sensitive sites of fishing needs special attention and motivation for future management. Disturbance in these sensitive geomorphic zones by intervention of the inhabitants of the Sunderbans will not only jeopardize the prawn seed population, but also will bring about severe damage to the water courses leading to bank erosion. Several causes of erosion in these areas include (i) felling of immature mangrove plants (ii) destruction of newly sapling and (iii) activities of man causing undercutting of banks.

10.14 Water Quality of Bidyadhari River

Water quality of adjoining wetland ecosystem of Bidyadhari river basin gradually has been worsened due to admixture of industrial effluents with domestic sewage released from the Kolkata metropolis as well as from the different industries of Kolkata east through different canals (Fig. 10.6). Pollutants from industrial sewage may cause biological magnification that ultimately leads to carcinogenic effect in the human body. Attempts are made to record some physico chemical parameters from six different stations of Bidyadhari River. It is now urgently needed for mapping of Bidyadhari river course through physico-chemical parameters to let the local people know the water quality of Bidyadhari River in different sectors for their benefit in application of water of Bidyadhari River in agriculture and aquaculture.

Bidyadhari River has a very circuitous tidal course in the district of North 24 Parganas, West Bengal. It has the special characteristic of carrying the tidal domestic sewage along with others which are thrown by a series of artificially made canal named Bagjola Khal, Bhangor kata khal, Central Lake canal, Krishnapur canal etc. All these canals now carry effluents from different industries. Bidyadhari River is tidal almost of its entire length within its upper limit up to Tehatta at the upstream.

Fig. 10.6 Sewage passes through the sluice gates of the canal into the Bidyadhari River

10.15 The River System

The Bidyadhari has a very circuitous tidal course. It begins as a drainage channel, called Nona Gang, a little north-east of Dogachia, which occupies a low interfluves between the Sunti and the Padma. It flows south past Jessore road and Barasat-Basirhat rail track. Just south of this rail track the Haroa gang meets the Nona gang on the right and the combined stream is known as Bidyadhari. The Haroa itself serves as drainage outlet of the extensive swamps to the east of the Dhapa bil around Rajarhat-Bishnupur (22°37′N and 88°30′E). The Haroa gang appears on the map as a continuation of the north-south flowing Sunti nadi through these bils. Further south (about 20 km) it receives the Bhangar kata khal (which drains the salt water lakes of East Calcutta) on the right bank and the Kulti (or Chaumuha gang) on the left bank. Below this confluence, the Bidyadhari flows south as the tidal Kumarjol gang. The Bhangar kata khal in its turn is linked with the Krishnapur canal, Circular canal and New-cut canal on the north eastern periphery of Calcutta. The sewage and storm water of Calcutta is canalized through two different channels into the Central lake canal, which empties into another branch of Bidyadhari south of Bamanghata (88°29′E and 22°31′N). The head water portion of this branch has largely silted up. It flows south east, receiving the Tolly's Nullah from the east and giving off the tidal Piyali River flowing south from its right bank before reaching Port Canning where it receives the other branch, i.e., Kumarjol gang referred earlier. This south flowing branch has other names – Khurti, Karats, Kuriabhanga and Kartoya. The united streams form the Matla River or rather estuary, which is navigable up to Canning.

Beginning in the Sunderbans, tidal Bidyadhari in the past flowed north-east past Haroa where it was known as Haroa gang, and then bent westwards, and was joined by the Nona khal. After this, it flowed south-westwards to the junction of the Beliaghata canal and Tolly's Nullah, and then southeast to Matla or Canning, where it was joined by the Karatoya and Atharonbanki rivers. The united streams formed the Matla River, which flowed south to the sea and was navigable by river steamer up to Canning (O'Malley 1914).

The portion of the Bidyadhari River near Kolkata, which at present serves as an outfall channel for the storm water and sewage of the city, has for some years past been silting up at a rapidly increasing rate. The acceleration of the silting process is attributed mainly to works in connection with local fisheries and to the reclamation of portions of the salt water lakes for rice cultivation, the effect being to decrease the spill of water from the river over the adjoining land and, consequently, to increase the deposit of silt in the river bed. Other contributory causes have been the construction of the Dhapa lock, the closing of tributaries in each of which the tide used to flow and ebb freely, and the canalization of the Bhangar khal.

10.15.1 Decline of the Bidyadhari River

It was observed during the span of 1901–1912, the bed of the Bidyadhari river had risen about 8 m in 8 years a 1.6 km below Bamanghata while cross-sectional area had been reduced from 715 sq m giving a mean rate of contraction of 45 m^2 and it was concluded from Hooghly – Bidyadhari Canal Enquiry Division that the Bidyadhari had a very short remaining lease of life, and that in 6 year time periodicity it would be useless as an outfall channel for the sewage of Calcutta unless remedial measures were taken. It was also suggested that the only practicable way of dealing with the situation was to canalize the channel of Bidyadhari from Dhapa to the off take of the Piali River, 15 km below Bamanghata, and the channel of the Piali River from its head to its outfall into the Matla River.

The Government of Bengal, acting on expert opinion, decided it would stop maintaining the river any further in 1928 and it was declared useless. The sewage of the then Calcutta metropolis was decided to be discharged into Kulti in 1935. Bidyadhari was absolutely dead without any chance of revival by 1942. The decline of Bidyadhari is due to various manmade and natural causes. Human intervention, especially, fishery and constructions of canals and embankments have accelerated the pace of degeneration of the river.

10.15.2 Present State of the Bidyadhari River

As a result the sewage-laden brackish water moves towards extreme upstream up to Tehatta -that lead to no use of river water for domestic as well as agricultural purposes. Water quality gradually worsens due to admixture of industrial effluents with the domestic sewage that pass through the different canals which ultimately finds its opening in the Bidyadhari main stream. It is observed that the industrial effluents play an important role in the adjoining wetland ecosystem as the discharges are being productively utilized in the aquaculture and are also in demand to irrigate the adjoining lowland agricultural efforts (Chattopadhyay et al. 2001). Present inclusion of industrial effluents in the domestic sewage may start the appearance of heavy metals in the water or nutrients that may cause biological magnification in both flora and fauna (Chattopadhyay et al. 2001). Accordingly attempts are made to study the physico-chemical and biological conditions of water flowing through Bidyadhari River. Temperature, Conductivity, DO, BOD, COD, TDS, TSS, pH, etc. are recorded from six different spots at the Bidyadhari river namely Haroa (1), Kulti (2), Ghusighata (3), Minakhan (4), Malancha (5) and Dhamakhali (6).

Biological magnification sometimes has harmful effects and even highly carcinogenic to human body. It is now urgently needed to map the upper and lower wetland limits in terms of physico-chemical analysis in Bidhyadhari river basin, which sustain the livelihood of the bulk of population of this area of Sunderbans. Geomorphic mapping will be helpful to the people surrounding the Bidhyadhari River through the sector wise utilization of river water for different purposes. Geomorphic mapping is further helpful by identifying the location area of sewage sludge accumulation. Removal of these sewage sludge from the river bed is not only helpful for easy navigation along the river, it will further act as the bio-degradable fertilizer free of cost to the poor farmer of grass root level.

10.16 Environmental Parameters of River Waters

Water samples from the six different spots (Table 10.3) named Haroa (1), Kulti (2), Ghusighata (3), Minakhan (4), Malancha (5) and Dhamakhali (6) from the north to south direction along the tidal stretch of Bidyadhari river (Fig. 10.6) were collected for the estimation of physico-chemical parameters of the water. Water of the above said six stations along the tidal river course of Bidyadhari were collected in polythene bottles of 1 l capacity for the estimation of pH, salinity, conductivity, TSS, TDS, Turbidity, BOD, COD, DO, oil and grease etc. (Ewing 1985).

The pH and oil & grease level (Table 10.3) of the surface water of Bidyadhari River at all sampling points have been observed to be well below the permissible limit (Fig. 10.7). This favorable pH (Table 10.3) as wells as the oil & grease concentration helps also the decomposition of organic matter by the microorganisms.

The TSS level in all the sampling points is well below the permissible limit (Table 10.3). This may be due to the settling of pollutants in all the canals flowing towards the Bidyadhari River. The comparatively higher level of TSS in case of Kulti may be due to unfavorable hydrodynamic condition in that region.

BOD level in all the sampling points is also below the permissible limit (Table 10.3 and Fig. 10.8). The surface water flowing through the Bidyadhari River carries some organic waste coming from the different man-made canals. The microorganisms are decomposing these organic matters in presence of sunlight and dissolved oxygen (DO) which are available abundantly in all those canals (Table 10.3 and Fig. 10.8). This is the reason of low level of BOD in those regions of Bidyadhari River.

Dhamakhali being nearer to sea shows the highest TDS level. The same is the case for Malancha, Minakhan, Ghusighata and Haroa (Table 10.3 and Fig. 10.9). The high level of TDS in case of Kulit may be due to contamination of wastewater of different industries with the surface water of the canal flowing towards Kulti.

The same is the case for COD level at different sampling points of Bidyadhari River except for the Kulti which shows the highest COD level (Table 10.3 and Fig. 10.8). This may be due to contamination of industrial wastewater in the surface water near Kulti region of Bidyadhari River.

Scientific information and environment analyses on the concentrations of different pollutants from the domestic sewage by the canal (Fig. 10.6) into the Bidyadhari River water and the resultant effect on bioaccumulation and biomagnifications are little known till date. The cumulative effects on the consumption of the agricultural and aquacultural products from the wetland areas adjacent to the Bidyadhari River in human body are yet to be studied. It is apparent from the present study that the wetland ecosystem adjacent to the Bidyadhari River plays an important role in pollution amelioration acting as a natural waste treatment plant (Das and Dutta 2004; Das et al. 2004). Sunlight and DO help in purifying the waste water which are admixtured to the Bidyadhari River water after being released by several man-made wastewater canals from the greater Kolkata. Feasibility test (Ghosh 1999) through analysis of environmental chemistry considering the result of major environment

Table 10.3 Analysis of water samples collected from different spots of Bidyadhari River

Sample no and location	pH	Salinity (ppt)	Turbidity (NTU)	TDS (ppm)	DO (mg/L)	TSS (mg/L)	BOD (mg/L)	COD (mg/L)	Oil and grease (mg/L)
B1 Haroa	7.25	3	17	460	9.2	0.0355	2.412	20	0.071
B2 Kulti	7.18	4	8	730	0.0	0.3500	4.824	92	0.112
B3 Ghusighata	7.02	4	9	590	0.4	0.0010	4.02	60	0.094
B4 Minakhan	7.14	5	12	670	2.2	0.0015	1.6	72	0.437
B5 Malancha	7.12	5	15	760	4.8	0.0130	4.824	52	0.243
B6 Dhamakhali	7.32	8	10	1,950	6.4	0.0100	1.608	68	0.695

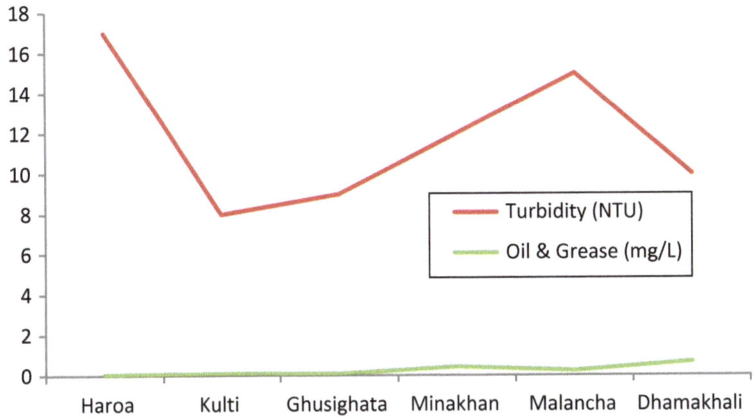

Fig. 10.7 Values of Turbidity and Oil & Grease at different spots of Bidyadhari River

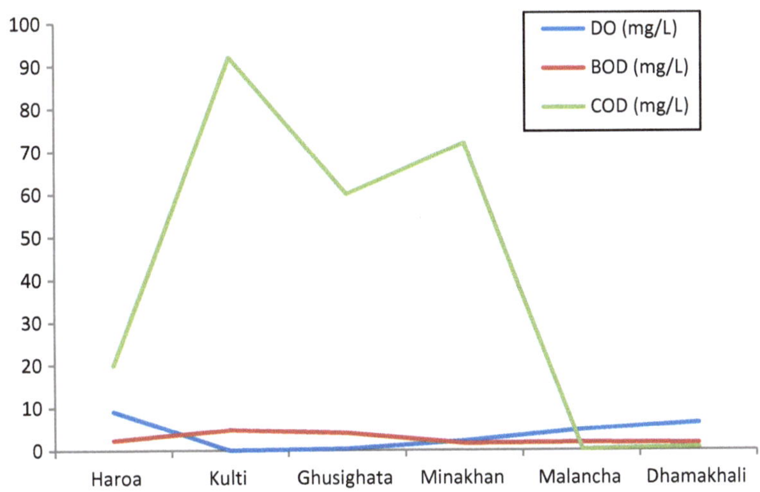

Fig. 10.8 Values of DO, BOD, COD at different stations of the Bidyadhari River

parameters reveals that the water quality of the Bidyadhari River is still tolerable for the existence of living creatures in the riverine aquatic environment. Only the sewage sludge present in the domestic sewage released by the canals causes impediment to inland navigation as well as marks sedimentation on the riverbed. Therefore, political will and social awareness are needed to restrict further pressure on both the products and the producers existing in the adjacent wetland areas of the Bidyadhari River.

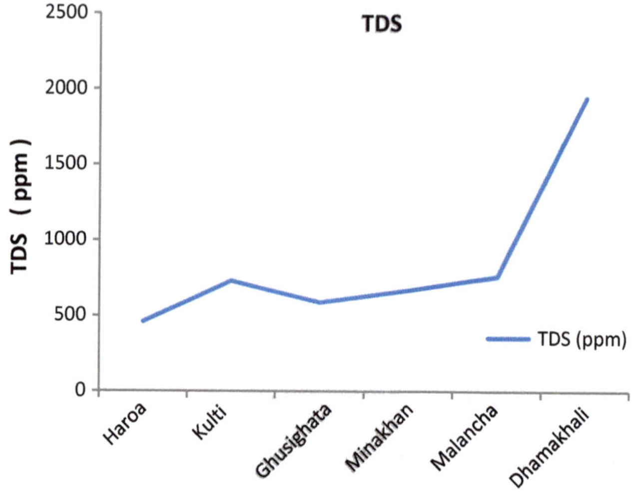

Fig. 10.9 Values of TDS at different sampling stations at Bidyadhari River

10.17 Man-Tiger Conflict and Tiger Straying

The mangrove swamp of the Sunderbans is famous for the habitat of the Royal Bengal tiger and their typical characteristics of man-eating habits. The tigers of the Sunderbans choose all the moving animals as their prey. They even prey upon the crocodiles lying in the river flood plains of Sunderbans. Straying of tigers depends on the physical as well as the biological factors. The age-old tigers are generally disabled for catching a moving wild prey in the forest environment rather than capturing domestic bovides in the locality as their part of menu. They never attack human being as their prey but attack if only the men are seen on their way.

10.17.1 Classification of Man Eaters

The mother tigress if disturbed by the honey collectors or the wood cutters using the passage through the tiger den, only attacks the intruders like honey collectors, fishermen or wood cutters for the protection of her cubs and these types of man eaters are considered as 'circumstantial man eaters'. Only 1 % of this category preferred attacking men as easily available and weak prey are considered as 'designed man eaters'. When this man eating habit is gradually acquired from the mother even by a male cub, they become an 'aggressive man eater' for whom the human beings are their normal preys. The wood cutters fishermen and the honey collectors enter into the tiger den as the trespassers are therefore responsible for the changing the habits of a tiger into a man-eater in the Sunderbans tiger reserves.

10.17.2 Occurrences for Man Eating

Royal Bengal tigers generally seen to attack in the mangrove swamps with higher saline zone have a positive correlation with the man eating of tigers. Maximum casualties are occurred in the zone of maximum salinity. Further maximum man eating takes place in the pre-monsoon time when salinity in the river waters is maximum. The tiger straying and man eating are therefore directly related with the salinity in the preferable pre-monsoon seasons.

Further if the honey collectors move frequently in the mother tiger den living with cubs in search of honey comb without the knowledge of the presence of tiger den, the tigers attacks the honey collectors for the protection of cubs and becomes a man eating tiger.

10.17.3 Measures for Minimization of Confrontation

Permission for harvesting of *Phoenix paludosa* is restricted as the continuous groves of this bushy species, the natural and preferred denizen of tiger den is disturbed no more. Electrified dummies of fishermen, honey collectors, wood cutters are introduced. The man eaters, Royal Bengal tigers are pretended after getting shock from those electrified dummies with their necks covering with electric wires having electricity with 230 V. The man eaters may avoid the men entering inside the forest for collection of honey, wood or fish catch as their prey being lured and pain from getting electric shocks from those human beings like dummies in the forest areas. Honey collectors and fishermen are advised by the forest officials to wear rubber made face mask on the rear side of the head to embarrass the man eaters confusing in taking decisions on side of attack as the Royal Bengal tigers of Sunderbans are habituated to attack from behind the forest goers like wood cutters, honey collectors and fishermen. Excavation of sweet water ponds inside the forest is another way of thoughts to stop the straying of tigers in search of fresh waters to the localities adjacent to the tiger reserves. These ponds in the buffer zone filled with fresh waters from the precipitation by the rain restrict the movement of the tigers within the forest.

At present the entry times of wood cutters, honey collectors inside the forest is restricted as tiger attacks are recorded in between 7 am to 9 am and 3 pm to 5 pm. Casualty in the night time is happened generally after 11 pm when a man eater selects a person sleeping on the boat, then catch him and jumps to the water with the prey holding firmly on his neck. Fences with nylon net and branches of mangrove trees covering those islands opposite to the villages restrict the tigers sometimes not to cross the rivers or creeks to the human habitation zones as they are specially adapted for swimming across the tidal rivers in the Sunderbans.

Table 10.4 Inventory of the litters scattered in the beach

Plastic packets	619	Doll's head	08
Plastic bottles	105	Plastic nine pins	16
Pieces of rope	113	Glue syringe	03
Shoes (not in pair)	52	Used condoms	09
Fishing net balls (punctured)	66	Small gas cylinder	04
Glass bottles	38	Foam pieces	19
Bottle tops	34	Drug syringe with intact needles	08
Pieces of plastic pipes	16	Worker's used materials	27
Jars	07	Plastic coat hanger	02
Broken foreign liqueur bottles	14	Toy (entire)	07
Fluorescent tubes	03	Half a toy	61
Light bulbs	01	Car floor mats	03
Food/drink cans	09	Asthma inhaler	01
Pop tops	02	Ampoules (empty)	38
Dice (Ludo)	01	Jar lids	07
Cigarette lighters (not in use)	17	Buoys: large	01
Cycle tire	01	Crates (bread, bottles)	05

10.18 Trash in the Coastal Sunderbans

Fraserganj, a natural solitary sea resort, only 1.2 km west from the well-known Bakkhali sea beach stands at the southern most ends of West Bengal where coastal landmass meets the Bay of Bengal. Fraserganj was so named after the name of Sir Andrew Fraser, the Bengal Governor during 1903–1908 who was more famous as the advisor of Lord Curzon for the division of the then undivided Bengal. Andrew Fraser took the project to set up a tourist spot for spending leisure time of the employees specially for the British nationals during the last year of his Governorship in Bengal in the year 1908.

That Fraserganj at the bosom of the Bay of Bengal is one of the most remote and lonely sea resorts situated almost 139 km from the nearest mega city Kolkata. Yet when a survey conducted on July 16, 2009, it was found over 1,000 pieces of trash in a 2 km stretch of this beach (Table 10.4). Anthropogenic causes have the major impact over the accumulation of trashes on the beach. Non-biodegradable trashes particularly the plastic end up in the sea coming from the lands (Kar 2007).

10.19 Summary

Sunderbans therefore, faces different man made problems a few of which is local and the source of the other one is Kolkata mega city that is far away from the Sunderbans. The prawn capture was started only from the late 1980s of the twentieth century with the introduction of highly profitable scientific brackish water prawn culture with direct or indirect support of multinational companies. This practice not only diminishing the tiger prawn population but also a large number of fingerlings and seeds of other prawn and fish species get trapped and vanishes from the nature. As a consequence, it affects negatively those animals including big fishes and even crocodiles who take these small creatures as food. The food chain will break up step by step and its deleterious effect may be more harmful in future. The prawn seed collection continues unhindered, it can decrease the density of pneumatophores and the biomass of epiphytic algae causing a change in habitat structure. As a result, trampling in the muddy river banks is a perennial problem.

Mangroves of the Sunderbans are under gradual increasing pressure and pollution due to the human activities because of rapid reclamation and encroachment in the coastal areas. Prawn seed collections, refuse disposal, sewage discharge, urban emissions, accidental spillage of toxic pollutants are significant anthropogenic inputs. Over exploitation activities in the Sunderbans have led to large scale mangrove area degradation despite the mangroves have immense direct and indirect uses. The mangrove forest zones have faced changed land use patterns in the form of conversions of mangroves forest lands to agricultural lands, aquaculture farms, and human settlements during the period of last two centuries. The ideal nursery grounds, grazing places and breeding sites for a large number of mangrove dwelling species have gradually been destroying for this human encroachment and interferences. The mangroves forest of the Sunderbans not only serve as the habitat of aquatic species but act as the coastal buffer rendering protection to the coastal region as well as the megacity Kolkata from the severe coastal upsurges and cyclonic storms.

The number of casualties due to tiger attack inside the forest has been gradually diminishing as almost half of the permit holders among the wood cutters, honey collectors and fishermen are happened to be involved in the collections of tiger prawn seeds (*Penaeus monodon*) in the river waters of the Sunderbans. The right awareness and the involvement of the village people after repeated meetings leads to the less casualties of the tiger straying in the villages. The man-tiger conflict and managing the straying of tigers in the localities are gradually understood through regular meetings held in the forest side villages by the Forest Protection Committee (FPC) and Eco-development Committee (EDC). These committees are formed by the joint membership of the forest officials and the villagers where the concept of eco-friendly relationship in between man and tiger are discussed. The forest officials make the villagers understand – Sunderbans will be saved only if tigers are saved which in turn necessitates a luxuriant growth of mangrove forest for their dwelling.

Trashes, in particular, in the tourist spot in the coastal Sunderbans pose hazards to the marine life specifically when the animals drown or strangle from being tangled in discarded or lost fishing gear or suffer and even die from swallowing plastic and other garbage. Plastic packets look like jelly fish are taken for feeding by the sharks and turtles.

References

Anonymous (1993) In: Banerjee BK, Singh H (eds) The shrimp fry by-catch in West Bengal. CIBA, Madras

Chattopadhyay B, Dutta S, Chatterjee A, Mukhopadhyay SK (2001) The environment impact of waste chromium of tannery agglomerates in the east Calcutta wet land ecosystem. J Soc Leather Tech Chem 84:94–100

Das GK, Bhattacharya A (1999a) Destruction of certain encountered shelfish and finish from the coastal waters of Sunderbans and its impact on the coastal ecosystem. J Indian Soc Coast Agric Res 17(1&2):216–218

Das GK, Bhattacharya A (1999b) Rampant exploitation of post-larval penaeid prawn (Penaeus mondon) from the salt-water course of the deltaic Sunderbans, West Bengal – a threat to the balance of nature. J Indian Soc Coast Agric Res 17(1&2):219–222

Das GK, Dutta S (2004) Studies on the impact of water quality on the adjoining wetland ecosystem of Bidyadhari River, West Bengal. Indian Sci Cruiser 18(4):16–21

Das GK, Dutta S, Sanyal SK (2004) Need for Geomorphic mapping in terms of physico-chemical analysis of the sewage fed Bidyadhari River carrying effluents from the greater Calcutta. J Indian Soc Coast Agric Res 22(1&2):49–51

Ewing GN (1985) Instrumental method of chemical analysis. McGrew Hill Book Company, New York

Ghosh P (1999) Proceedings of habitat II conference, Calcutta

Kar D (2007) Public access and importance of offers to dedicate on California's coastline. J Coast Res 23(2):472–486

O'Malley LSS (1914) Bengal District Gazetteers, 24 Parganas. The Bengal Secretariat Book Depot, Calcutta

Chapter 11
Changing Environmental Scenario

Abstract Various morphodynamics factors together with anthropogenic interferences are bringing in substantial changes in the unique ecosystem of Sunderbans Biosphere Reserve. Significant rise in sea level, high rate of erosion and accretion in the deltaic system, change of salinity and temperature of river water, change of river course causing impeded fresh water flow in the estuarine domain, over exploitation of mangrove vegetation which acts as a buffer to cyclonic storms and construction of embankments along the islands are some of the natural and human factors responsible for causing ecological imbalance and depletion of mangrove vegetation with its diverse floral and faunal habitats. A comprehensive multidisciplinary approach for management of Sunderbans as well as awareness among the Sunderbans dwellers about importance of mangrove and its conservation is the need of the hour.

Keywords Sunderbans • Mangroves • Salinity • Tidal flow • Estuarine ecosystem • Embankment • Anthropogenic activities

Sunderbans mangrove is a natural phenomenon and an integral part of the dynamic Ganges-Brahmaputra deltaic tidal plain. Natural phenomena do not depend upon the subjective desires of the human race as because nature is guided by its own laws, the laws of nature. Man can definitely govern nature only a limited extent. It can be safely concluded that sooner this limitation is understood, safer will be survival in this planet and that is true for managing the Sunderbans.

Coastal urbanizations and changes in local hydrology, regional and global processes are the most important contributory factors attributable to changes in mangroves of Sunderbans. Sunderbans with 9,630 km^2 entire areas of 100 years back is at present left with 4,266.6 km^2 area covered with the mangrove vegetations and the rest has already been converted to various land use pattern. The Sunderbans mangroves, the coastal guard, at present has gradually been made ineffective, one of the most productive ecosystems rendered unproductive and the only mangrove tiger land of the world was exposed to threats of extinction.

The rate of resulting net sea level rise is 3.1 mm per year computed at Sagar Island, the biggest delta of the Sunderbans (IPCC 2007). This leads to the adverse impact on the climate and environment of the Sunderbans such as flooding of

G.K. Das, *Estuarine Morphodynamics of the Sunderbans*,
Coastal Research Library 11, DOI 10.1007/978-3-319-11343-2_11

low-lying deltas, retreat of shorelines, salinizations and acidification of soils etc. The salinity of waters along the coastal tract of Sunderbans has been decreasing gradually due to dilution of seawater. Temperature of the flood-tidal river waters of Sunderbans is recorded higher than that of the ambient temperature during winter time. Silt particles undergone mixing with water and the resultant heat of mixing as a result of the kinetic energy from those vortexes may cause the water temperature warmer than the ambient temperature.

Mangroves vegetations have a strong relation with the fresh water inundation. Reduction of head water discharge has damaged the vegetation dynamics and changed the land scape pattern in the Sunderbans. This situation has also the negative impacts on the environmental, ecological, geological and geomorphologic set ups. The impact over the soils causes the degradation of the surface organic matters and as a result of that the growth of mangroves has become stunted with gradual decreasing of soil fertility. Ultimately the mangrove ecosystem has been damaging as a whole with gradual degradation of water quality.

In the estuarine Sunderbans, growth of mangrove vegetations and their productivity are affected by the salinity. Climate, hydrology, topography and tidal flooding control the salinity of river waters in this estuarine system. Salinity gradients may be considered as one of the most important factors that can influence the distribution of plant species in many instances in the Sunderbans. In lower saline areas the mangroves are seen more luxuriant. Chronic high salinity is detrimental to the mangroves. The mangrove plants are seen to have some tolerance of salinity. Hypersalinity stunts plant growths, reduces biomass in hydroponically and causes denaturing of buds.

Responses of the mangrove floral community to the soil conditions are directly correlated. Nutrient availability may limit growth and production of mangroves ecosystems. Naturally one of the most important factors appears to be nutrient concentration. Variation in nutrient concentrations may change competitive balances and affect species distributions. Nutrient pulses can create immediate and impressive changes in the mangrove vegetations. High nutrient concentrations and low salinity during monsoon period produce rapid growth in the mangrove saplings.

The topography of Sunderbans is very low and flat. The deltaic region is the most active one, and continuous processes of accretion and erosion are going on in this Sunderbans. Sagging down, undercutting and collapsing of rivers in several areas can be supposed to be the evidences of erosion followed by differential subsidence and filling up of sediments. The high rate of erosion is evidenced from the relics of the sea wall exposed in the intertidal beach zone and estimated to be 9.8 m/year during the period 1930–1970 and 8.6 m/year during 1971–1995 in the coastal areas of the Sunderbans. The northeastern, southeastern and southwestern sides of Sagar Island are facing vigorous erosion due to the concerted acts of various natural processes and anthropogenic activities. The erosion rate from 1996 to 1999 was calculated to be 5.47 m/year (Gopinath and Seralathan 2005). The part of the destabilized dune has been advancing inland at the rate of 17 m/year. Several wave and tidal current related erosional features exhibit a special sequences from the seaward side and include furrow and ridge structure, mud balls and mud mounds.

The erosional features like furrow and ridge structures, mud balls and mud mounds help in the understanding the morpho-dynamic processes that lead to the landward displacement of the coastal Sunderbans which reached its present position of about 5,000 years BP. *Amalda ampla*, a gastropod mollusk occurred in the coastal beach of the Sunderbans may be considered as biological indicator for the determination of the beach erosion (Das 2009).

Large scale construction of concrete embankment all over the 54 islands out of total 102 islands of the Sunderbans which is the dwelling place for about five million people has the major impact over the fragile ecology of this mangrove lands. The concrete embankment causes the barrier to the unique mangrove dwelling fauna at the natural levee and river flood plains resulting detrimental effects on the mangrove ecosystems. The absence of a particular fauna leads to a definite gap in the food chain and ultimately the structural frame work of the food web becomes disturbed. The river banks of the Sunderbans is not only the habitats of hundreds of non-chordates like pea crabs, fiddler crabs, sea pens or sea anemones, but is the denizen and breeding grounds of the birds like king fishers and sea gulls. The concrete embankments are not feasible in protecting the embankments that are seen eroded away by the tidal actions in the Mahishani Island of the Namkhana block. Polypropylene sheets used for the concretization of the embankment will be harmful to the estuarine environment as the polypropylene is not a bio-degradable ingredient. As a result the construction of the concrete embankments using polypropylene intervene the habitat structure of the mangrove fauna resulting detrimental impact over this dynamic ecosystem.

The highly specialized mangroves ecosystem of Sunderbans basically can be treated as an estuarine ecosystem supporting a large number of animal communities living either entire life cycles within the mangroves or visiting the mangrove swamps and waters to feed or to breed. The extreme condition of this estuarine environment has resulted the vulnerable effects for both the flora and fauna even for their occurrences in this fragile ecosystem. Expansion of mangrove swamp to the salt marsh is the clear indicator of changing trends of the scenario of the mangrove vegetations. Increased tidal amplitudes and changes in riverine nutrient and sediment loads could lead to the expansion of mangroves into salt marsh areas.

Islands emerged in the estuarine river mouths and coastal areas of the Sunderbans are gradually disappearing because of the climatic changes and sea level rise. But the tale of Jambu dwip is quite different. Human stress for fish drying purposes and their habitats are the vital reasons for the exploitation of mangrove resources in the river mouth island, Jambu dwip that stands in the estuary of Hooghly River along the coast of the Bay of Bengal. The anthropogenic interference including clearance of mangroves forest to organize the fish drying operations by the merchants and thousands of labourers has accelerated erosion in the southern part of Jambu dwip.

Deposit feeders colonize preferentially during the post monsoon season coinciding with the increasing amount of detritus in the sediments. The density of population of these deposit feeders is not same in all the seasons. The abundance and distribution of population depends not only on the physico-chemical parameters like DO, salinity, temperature, pH of soil and water but on the quantity of detritus present in the habitat although living organisms themselves organize very important biotic complex in the structure of biotic community. Environmental heterogeneity is observed in a species rich area constituted with the detritus during dry season than that of a species poor area in the mangrove lands of Sunderbans.

A large number of encountered species of fin fishes and shell fishes are suffering a severe loss in population because of random exploitation in addition to the damage caused to the prawn seeds. Continuous prawn seed collection can decrease the density of pneumatophores and the biomass of epiphytic algae as trampling is a perennial problem which affects the habitat structure of mangrove plants as well as aquatic organisms. This, in turn, adds to a tremendous threat to the balance of the coastal ecosystem.

Environmental conditions along with changes in weather conditions and associated modifications due to climatic changes are strongly associated with the eco-tourism (Coombes et al. 2009). Increased temperature as a result of global warming will lead to a greater number of tourists not visiting the Sunderbans round the year. In particular, currently tourists restrict their visits to Sunderbans for eco-tourism only for about 4–5 months during winter season in this only tiger land of mangrove forest in the world.

These changes are further influenced by the changes in river course, impeded fresh water flow, climate variability and sea level changes. All these deleterious operations are currently continuing in most areas of Sunderbans, which, in turn, during the recent decades, have been bringing about a serious change in the geomorphology and hydrodynamics of the watercourses together with an obvious change in the nature of nutrient recycling. As a result of these morphodynamics processes as well as anthropogenic interferences like over exploitation on harvesting mangrove vegetations have led to the large scale mangrove areas degradation. Further, mangroves have also been lost through natural causes including erosion and severe siltation as a result of sedimentation. The trend of salinity decline, increase of temperature of river waters, non-uniformity of bottom topography, increasing rate of sedimentation in the river bed and mid channel bar formations; collapsing of bank materials, random collection of prawn seeds resulting damage of ecological balance, ingression of salinity into the inland water bodies leads to the present dwindling environmental scenario of the Sunderbans.

This dwindling environmental scenario of the Sunderbans by natural phenomena as well as by the anthropogenic interferences affecting the mangroves and mangrove ecosystems are still the causes of concern for the nature lovers, natural resource managers and scientists working in the field of conservation. Integrated management of this unique ecosystem through various schemes and implementation of positive steps are the need of the hour at the governmental level for conservation and development of the Sunderbans, as well as for the large number of forest fringe villagers who depend on the mangrove forests for fishing, honey collection, fuel and fodder collection etc. for their livelihood. Further, no governmental efforts to conserve and create natural resources can be successful and sustainable unless the people concerned actively participate in planning and management of the resources. The best effort to combat the exploitation of mangrove forest resources is to rekindle the sense of awareness among Sunderbans forest dwellers about the importance of mangrove and its significance as a natural protection from natural calamities. At the same time, alternate livelihood options are also to be created for the Sunderbans villagers.

Sunderbans, selected as the World Heritage Preservation Site, accounts for the biggest mangrove ecosystem in India, whose management and conservation is vital for existence and sustainable development of Calcutta and coastal West Bengal. While development aims to achieve human goals largely through the use of the biospheres, conservation aims to achieve them by ensuring that such use can continue.

References

Coombes EG, Jones AP, Sutherland WJ (2009) The implications of climate change on coastal visitor numbers: a regional analysis. J Coast Res 25(4):981–990

Das GK (2009) Beach processes of Tidal Islands of Hugli Estuary, West Bengal. Geogr Rev India 71(3):240–248

Gopinath G, Seralathan P (2005) Rapid erosion of the coast of Sagar Island, W.B. India. Environ Geol 48:1058–1067

IPCC (2007) Climate change-2007. Impacts, adaptations and vulnerability. Contribution of Working Group II to the fourth assessment report of the Intergovernmental Panel on Climate Change. Cambridge University Press, Cambridge

References

Index

G.K. Das, *Estuarine Morphodynamics of the Sunderbans*,
Coastal Research Library 11, DOI 10.1007/978-3-319-11343-2

MIX
Papier aus verantwortungsvollen Quellen
Paper from responsible sources
FSC® C105338

If you have any concerns about our products,
you can contact us on
ProductSafety@springernature.com

In case Publisher is established outside the EU,
the EU authorized representative is:
Springer Nature Customer Service Center GmbH
Europaplatz 3, 69115 Heidelberg, Germany

Printed by Libri Plureos GmbH
in Hamburg, Germany